The Iowa River Valley near the Amana Colonies beckons tourists today, more than a century after it attracted settlers to the frontier.

Photo courtesy of Amana Colonies

MAKE NO SMALL PLANS
A COOPERATIVE REVIVAL FOR RURAL AMERICA

LEE EGERSTROM

LONE OAK PRESS, LTD.

MAKE NO SMALL PLANS

A COOPERATIVE REVIVAL FOR RURAL AMERICA

BY

LEE EGERSTROM

PUBLISHED

BY

LONE OAK PRESS, LTD.

304 11TH AVENUE SOUTHEAST

ROCHESTER, MINNESOTA 55904–7221

507–280–6557   FAX 507–280–6260

First Edition

ISBN NUMBER 1–883477–04–2

LIBRARY OF CONGRESS CARD CATALOG NUMBER 94–072991

Cover photo: farmer–members of James River Farmers Elevator at Groton, SD, pitched in with the community's townspeople to build a nine-hole golf course on the edge of Groton. The golf course venture helped strengthen the broader community, bringing town and country together in collaborative action, And while this venture was taken for quality of life purposes, the golf course is credited with raising Groton property values by 25 percent. *See:* Aeilts, David. "Cooperation Par for the Course." AgriVisions Magazine. Harvest States Cooperatives. September/October, 1993.

For Stanley and Dorothy,

    who knew the beauty and strength
       of rural America;
and Clyde and Louise,
    who encouraged me to write about it.

Bob Bergland retired in 1994 from a long public service career dedicated to agriculture and rural America. He and wife Helen Bergland returned to their original hometown at Roseau, Minnesota, near the Canadian border, where a son–in–law and daughter operate the Bergland family farm.

He was Secretary of Agriculture from 1977 to 1981. More recently, he served as president of the National Rural Electric Cooperatives Association in Washington, D.C., from 1981 to 1994. And he was previously elected to four terms in Congress from Minnesota, serving from 1971 until 1977 when he resigned to become Secretary of Agriculture.

FOREWORD   BOB BERGLAND

*A new rural America is starting to
take shape, rising from the ruins of communities
that were no longer needed to serve the needs of
traditional agriculture.*

In that one sentence, Lee Egerstrom has summarized both the tragedy and the hope of rural America. It is tragic when change doesn't treat people or communities kindly. But there is hope as well. Technology is causing change; we can use it and adapt it to make an even better rural America.

I know change doesn't come easy. My dad bought one of the first tractors in my area. He wasn't popular with people who had a vested interest in the horse trade. I've seen opposition rise up against every technological change since the tractor because each change has economic consequences.

More than 20 years ago, campaigning for a seat in Congress, I spent much of my time in towns that were sick, if not dying. New technologies were being used and world markets responded to the change. The generation of farmers and merchants from that era has accelerated the flight of rural population to larger cities. Schools have closed or, at best, consolidated beyond recognition, and rural hospitals felt the pinch of evaporating federal subsidies. For every business that was still operating on main street, several storefronts had boarded up.

These trends haven't reversed. If anything, they are gaining momentum.

"Conventional wisdom" has called for tinkering with New Deal approaches to rural economic problems. They had, after all, been fairly effective in slowing the hemorrhage since the Great Depression. Adjust federal subsidies for commodities, "feed the hungry," inject more federal funding to "infrastructure."

With the help of a few bad crops in other parts of the world, these programs did slow the decline. But it was still a hemorrhage, and it could not be stopped with a whole box full of legislative bandages.

Needless to say, things have gotten worse. As economic conditions continue to deteriorate in rural America, the response in Washington has gone through three stages of evolution. First, there was a seemingly endless flow of federal dollars. Then an unwillingness to send what rural Americans considered necessary. Finally, the nation was no longer able to meet the demands. Budget deficits don't allow a massive infusion of federal dollars into the far reaches of rural America. It hardly matters. International trade agreements now prohibit what were often standard U.S. reactions to world market problems and surpluses.

ဢ၆ဢ၆ဢ၆

Lee Egerstrom and I are just a couple of farm kids who happened to get pretty good jobs in town. From different perspectives, we've followed the same decline and shared the same pain over changes we've seen. On the streets and in the fields of rural America we have heard the same voices – sometimes complaining, sometimes explaining – forcefully documenting the decline. At world food and agriculture conferences we have listened to the same frustrations, the same "solutions" and, occasionally, the same arrogance of international policy makers.

The journalist, Lee Egerstrom, and I, a sometime policy maker, have come to the same conclusion. The "answer" to the problems of rural America is going to come from rural America, not cities and policy centers like Washington, Ottawa, Brussels and Geneva. From the bottom up, not from the top down.

We have seen exciting examples of rural America taking charge of its own future. Forward looking communities and farmers are taking the initiatives necessary to spur a true revitalization of the rural economy and they are doing it by simply finding their own roots. No, they are not attempting to turn back the clock to a rather vague Golden Age in the countryside. They are remembering and rekindling the cooperative spirit of our grandparents and great–grandparents as they search for solutions to frontier problems with markets and services. Put another way, they are adjusting to change and finding ways to pool resources and use new technologies.

ဢ၆ဢ၆ဢ၆

This book reminds me of a new promotional tool that has become popular in the countryside. Grain elevators and rural merchants have developed it as a gift item to supplement the seed corn hat. It is old technology – it is the "Big Buster."

It's just a two–foot section of a 2" by 4" with the less–than–serious purpose of giving your point of view more weight in an argument or gaining someone's undivided attention.

Sometimes when we dream of the way things used to be, we do need to be hit over the head to bring us back to reality. <u>Make No Small Plans</u> is a Big Buster for rural America.

A favorite excuse most of us use when discussing the "plight of the family farmer" or the "decay of rural America" is to attribute everything to "forces beyond our control." We try to personalize those forces and look for someone to blame. There are no villains here! I tell people they must stop looking for someone to blame and start looking for workable solutions.

Technology is changing us. If we change with it, and make wise business decisions, there are no forces beyond our control. <u>Make No Small Plans</u> is a call to action and a blueprint for rural America to take back control of those forces – and get the job done. If we change, the future doesn't need to be tragic. There can be a golden age ahead for agriculture and for all of rural America.

MAKE NO LITTLE PLANS;
THEY HAVE NOT
THE POWER TO STIR
MEN'S SOULS

Howard Cowden's business motto for building Farmland Industries into a modern agribusiness and petroleum giant is permanently engraved in the cooperative's boardroom wall.

Photo by Carolyn Riddle

PREFACE

"Make no little plans. They have not the power to stir men's souls."
Howard A. Cowden,
founder, and president of
Farmland Industries,
1929–1961 [P–1]

The quote is engraved in the Boardroom woodwork above the directors table. It is partly to remember the bigger–than–life founder of Farmland Industries, and partly to inspire new generations of leaders at the large Kansas City farmers and consumers cooperative. Howard Cowden used it time and again in speeches exhorting farmers and rural townspeople in Missouri, Kansas, Oklahoma and Iowa to band together in cooperative action. His call must echo again across America's Heartland if rural America is to revive and survive in the Twenty–First Century.

I would go further: It should inspire creative thinking among leaders of American labor and urban communities as well. The time has come to find new structures for the ownership of production facilities and to build job security through employee equity stakes in plants, urban and rural. The Basque people in northern Spain provide models through their Mondragon industrial cooperatives. Similar ventures are being started on the West Bank and Gaza, and at former state–owned factories and farms in Russia and Eastern Europe. Workers at a cheese factory in Wisconsin have likewise taken over ownership before their plant could be closed, and workers in Canada are preserving their jobs by buying their plants, often with the help of community development programs. But this is beyond the scope of this book.

Instead, Make No Small Plans – a slight play on Mr. Cowden's rallying cry – is intended as a fresh call for cooperative action in rural America at the start of the new century. Failure to act will have catastrophic consequences for residents of rural towns and farmsteads. The world economy has changed dramatically. Governments have lost both the legal authority to intervene in farm markets and arbitrarily raise farm incomes by raising prices, and the budgetary means to do so.

International trade agreements ban the former; political support and government budget priorities limit the latter.

Tom Webb, a Washington correspondent for Knight–Ridder Newspapers, explained this well in a February, 1994, dispatch to his newspapers:

> U.S. Secretary of Agriculture Mike Espy has seen the future, and rural America may not like his vision.
> "I have seen the handwriting on the wall ... U.S. budget support to agriculture will continue to decline. We can scream, we can curse, we can lambaste, and sometimes even cause some delay in the inevitable. But the fact is that U.S. budget support for agriculture will continue to decline."

There is bipartisan consensus on this, Mr. Webb noted. He quoted Senator Bob Dole, the Senate Republican leader from Kansas, on the declining farm subsidies. "Our future lies in more, unfettered trade," he said. And Representative Pat Roberts, the Kansan who serves as ranking Republican on the House Agriculture Committee, stated the harsh realities about future farm programs: "Every time a new farm bill is written, people say we're at a crossroads, but man, this time it's really true. We're going to be facing very severe budget restrictions. Ag is on the chopping block." [P–2]

This book accepts those assessments of the future of American farm programs. It also accepts the premise that momentum towards open markets and freer trade is irreversible: retreat from trade agreements would threaten the global economy and no major industrialized country should risk protectionist policies to support its farmers, laborers or any other class of workers and producers. Another premise of this book holds great sociological importance for rural America: Modern agriculture, with its new sciences and new technologies, no longer needs many people to produce food and fiber crops. The trends are in place and should be recognizable to all. A declining farm population in turn reduces the need for rural retail services and for rural communities to serve as senior citizens centers. In short, we have more rural communities marking spots on maps than are needed to serve people, based on current economic and demographic trends. [See Chapter Note P-A]

By accepting these premises, one must search for alternative actions. There are many; at least as many as there are creative people living and working in rural America. Sooner or later, one alternative becomes inescapable. Rural communities, townspeople, workers, farmers, foresters and tourism and hospitality people together must explore starting a new generation of cooperatives and new cooperative ventures.

It is starting to happen. There were more than $1.1 billion in new plants and cooperative ventures being planned in rural Minnesota and North Dakota by 1994, and state cooperative leaders expected the number and investment in such projects would double by the end of the decade. But the creation of what academic researchers call "new generation" cooperatives is fairly limited to the Upper Midwest states and a few isolated ventures in the Pacific Northwest, says Mike Cook, the economist and cooperative management specialist at the University of Missouri at Columbia. It hadn't started to spread to rural areas outside these Northern agrarian states by the mid–1990s. But, adds Professor Cook, they will. And they should.

Cooperatives are the most efficient vehicles for developing value–added business ideas and raising community capital to turn ideas into action. These businesses raise the value of area raw materials, such as farm commodities, and give producers a portion of processing profits at times when commodity prices are low and producer incomes are inadequate for family and farm expenses. They also provide jobs and gainful employment in rural communities for people no longer associated with the land. And those jobs, in turn, give communities a need for retail services, vibrant schools, churches and community services.

Yes, some fortunate communities will have entrepreneurs step forward with great ideas and the capital to launch new ventures in their hometowns. Other rural communities will be transformed into satellite manufacturing communities by virtue of proximity to major urban areas. They gain jobs and factories now spilling out from the higher land costs and space limitations of metro centers. But most rural communities can't count on fortuitous circumstances.

It's time to reach out and look for ways to band together, cooperatively, just as Howard Cowden and his contemporaries did during the Great Depression. Doing so won't be easy, as noted by Gilbert Fite, the University of Georgia historian. He recalls extraordinary moments when farm women in Kansas and Missouri raided household money from cookie jars to build what has become Farmland Industries. In his corporate biography, Beyond the Fence Rows, Professor Fite recalls that farmers and consumers in rural communities grasped Mr. Cowden's arguments that they needed to work cooperatively to improve their standards of living. [P-3] That message lost momentum along the way as agriculture and the U.S. economy grew and prospered. It is starting to be understood again.

The Fargo Forum newspaper in North Dakota reported in 1994 that farmers around the wheat farming community of Drayton sold 500,000 shares in a new value–added wheat products venture, Drayton Grain Processors Inc., on the first day the prospective co–op's equity was put on the market. [P–4] It was enough to start the cooperative although equity sales would continue.

Cookie jars have gotten larger. The purpose for raiding them remains the same. About the time the Fargo Forum reported on the Drayton area farmers, veteran Minnesota business editor and writer Dave Beal told his readers how North Dakota, South Dakota and Minnesota were vying for a $245 million corn processing plant. It will be built jointly by a new cooperative formed by corn farmers in those three states and two existing sugar cooperatives based in Minnesota and North Dakota. [P–5] The new plant would allow neighboring corn farmers and sugar beet farmers to jointly market about $1 billion worth of their fructose sweetener and sugar. And it would also open the way to use those sweeteners in more advanced food and soft drink processing that may be started by the farmers in the future. Mr. Beal explained the broader economic benefits of this corn plant in his column:

"The plant will provide jobs for about 900 construction workers. About 150 people will work there in a range of jobs, for average annual wages and benefits of $47,000. That's only $7 million, but Golden Growers President James Horvath [P–6] estimates that corn purchases, taxes, profits and various spinoff effects will boost the project's overall economic impact to more than $200 million a year." [P–7]

Such cooperative ingenuity revives and builds rural communities. But it is only happening where there is local leadership and a tradition of community activism. Still better examples of the new generation of co–ops and cooperative ventures can be found abroad. In Northern Europe, where Mr. Cowden gained inspiration on visits in the 1920s and 1930s, and in Japan and the newly independent states of the former Communist bloc, farmers are banding together to form new generations of cooperatives. They either process and manufacture products under contract with giant marketing companies, or they do so in joint venture businesses they've formed with successful marketing companies. In Russia and other Eastern European countries, new co–ops are being formed to produce and move food from farm to supermarket store, filling voids in the food distribution system left by the collapse of Communist governments.

The new enterprises are following models that are largely the invention of Northern Europe, from where most American cooperative businesses also have roots. For example, most domestically produced foods in Denmark today are handled from farm to table by five farmer–owned cooperatives. [P–8] Shared earnings from different rungs on the food ladder have allowed Danish farmers to continue operating comparatively small but sophisticated farms, keeping people on the land and rural communities strong. Working cooperatively, these Danish farmers have the world's most advanced hog industry and lead the world in pork exports. [P–9] Cooperating groups of Dutch, German and French farmers are also ahead of their American colleagues in forming joint venture business ties with international companies, and they are forming their own large entities to market products worldwide. [P–10]

These models must be given fresh looks. The American prairies are a frontier again. The people of the prairies must adapt to changing world economics and technologies, and rebuild businesses for the Twenty–First Century. The alternative is to surrender, to demolish redundant townsites so land can be cleared for corn, wheat and soybean farms that ship raw grains and oilseeds to other communities and countries for processing into more valuable products.

This book, then, is a call to action by a journalist who has monitored and chronicled political and economic changes for the past three decades. The first chapter offers an historical and cultural perspective for why cooperatives can revive rural America, especially in the old Northwest states. The second and third chapters recall personal experiences in both studying and tripping over irreversible changes in agriculture and world markets. The middle three chapters explore successful cooperatives in the Netherlands, here at home, and alternative American communities that continue to succeed by maintaining a strong cooperative spirit. The final three chapters look ahead at the technology and knowledge transfers to other lands that should be utilized at home, at the tools American states and communities are using to promote economic development, and at the creative explosion of new cooperative ventures now being started that hold promise for rural America in the Twenty–First Century.

PREFACE FOOTNOTES

1. Crank, Nellie Mae. In an interview for this book, Ms. Crank recalled Howard Cowden using the phrase often in the late 1920s and early 1930s, when references to men and mankind didn't have gender significance. Moreover, the words weren't his own invention. Promoters of the Chicago World's Fair had coined the phrase to justify their grandiose plans, she says. Nevertheless, they served Mr. Cowden

well since there was nothing timid about his plans for a consumer petroleum cooperative.

2. Webb, Tom. Knight–Ridder Newspapers, Washington, D.C. From news dispatch of Feb. 17, 1994.

3. Fite, Gilbert C. Beyond the Fence Rows. University of Missouri Press. Columbia (Mo.) and London. 1978.

4. Fargo Forum. "Drayton Grain Processors exceed requirements." Fargo, North Dakota. June 21, 1994.

5. Beal, Dave. "Three states vying for sweet business." St. Paul Pioneer Press, St. Paul, Minnesota. June 24, 1994.

6. Northern Corn Processors has since reorganized and changed its names. The joint venture processing company is now called ProGold Limited Liability Co. and the new corn growers' cooperative is now called Golden Growers Cooperative.

7. Beal. "Three states vying for sweet business."

8. Business Denmark 1993. Business Denmark Publications ApS., Copenhagen; and Denmark Review. February, 1994. Royal Danish Ministry of Foreign Affairs. Copenhagen; and Hansen, Otto Ditlev and Jorgensen, Erik Juul. Food Technology in a Modern Food System – The Case of Denmark. A monograph published by The Institute for Food Studies & Agroindustrial Development. 1993. Copenhagen.

9. Most American consumers are probably familiar with the Plumrose brand. Danish cooperatives process and package its hams and meat products, but the brand has belonged to East Asiatic Company, one of Denmark's oldest and largest public companies. At the time of this writing, EAC had investment bankers seeking buyers for its Plumrose subsidiary. Another brand familiar to many American consumers is Ess–Food. It is a trading company joint venture business owned by Danish meat cooperatives.

10. Egerstrom, Lee. "Rediscovering Cooperation." A monograph published by the Minnesota Association of Cooperation. St. Paul, Minnesota. 1993; and Fairbairn, Brett. University of Saskatchewan, Saskatoon, Saskatchewan. From research in Germany for a forthcoming history of German cooperatives to be published in Canada, 1995.

The following graphics show peak population periods for counties in the Upper Midwest, prepared by Dr. Randy Cantrell, demographic researcher at the Hubert H. Humphrey Institute of Public Affairs, University of Minnesota.

Many of the counties shown as having peak populations prior to 1940 actually reached their peaks in the 1870 census, he says. Its significance can be interpreted in different ways. But among them, it does show that no single set of federal farm policies can restore farm prosperity that would sustain large populations on farms and in neighboring towns. Indeed, the graphics should show rural community leaders that they are part of a frontier that either needs development or public and private plans for orderly disintegration.

POPULATION CHANGE for NORTHWESTERN COUNTIES
1980 - 1990

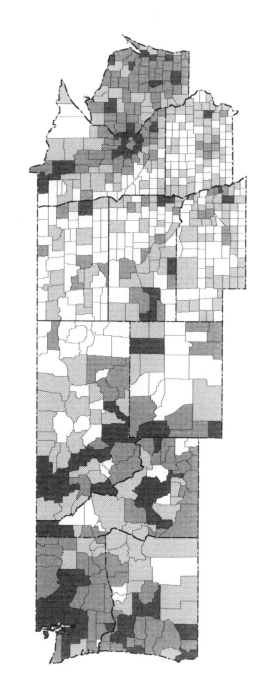

Rate of Change

Rapid Growth (>10%)   (75)
Slow Growth (<19%)   (161)
Slow Loss (<10%)   (213)
Rapid Loss (>10%)   (220)

Source: U.S. Census of Population

YEAR of PEAK COUNTY POPULATION

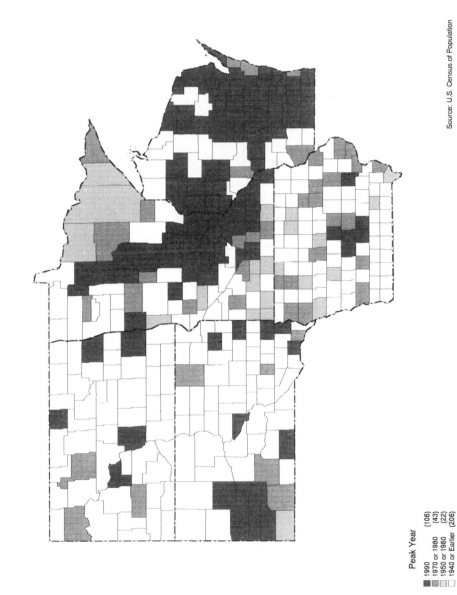

Source: U.S. Census of Population

Peak Year

■ 1990	(106)
■ 1970 or 1980	(43)
▨ 1950 or 1960	(22)
▨ 1940 or Earlier	(206)

19

TABLE OF CONTENTS

" ... Where can I buy me some sick steers?".
White House Photo

Chapter 1
Agrarian Myths &
The Northwest Culture

"Anyone who compares US and EC farm policy can see that, despite the differences of structure and of instruments, we have a lot of similar problems.

"Another thing we have in common is that, despite the decline in numbers employed in farming, our farm groups are still very vocal and influential with both politicians and voters. I found another reminder of the importance of the farm vote in America in that recent copy of <u>Newsweek</u>, where the first picture in the "US affairs" section was of Jesse Jackson trying to milk a cow."

Frans Andreissen, vice president of the European Communities. From an address given July 18, 1987, at the European–American Journalists' Conference, Copenhagen, Denmark.

"What you farmers need to do is raise less corn and more hell!"
Mary Elizabeth Lease,
Campaign organizer, Kansas People's Party, 1890.

Buried in nearly every American's heart is an emotional linkage to the land. It may be fond memories of visits to grandparents' farms, or cultural, ethnic ties kept alive by family traditions. It may be nothing more than a wistful attraction developed during peaceful drives into the countryside on vacations and weekend outings. Whatever the source, urban people and politicians won't let the linkage die. They keep repeating and amending centuries' old myths about rural people and farms, even when they are willing to let farm and rural policies wither and fade.

America has created a "noble peasant," the family farmer, to share rural America with the "noble savage," the Native American. The latter is honored in American literature and culture, but otherwise largely ignored. The peasant, too, is honored in literature and movies, and is starting to be ignored. It is logical that rural–based literary works are popular in rural areas, where they have regional importance. Their popularity beyond the heartland, however, owes much to the mythical notion that what is wrong with America comes from a nation's people losing their roots in the land.

If people living in rural communities are to band together to salvage what they have that is worth keeping, develop their resources now under–utilized, and build a culture worth preserving into the next century, the difference between agrarian myths and agrarian cultures must be understood. The former must be shed; the latter should be nourished and taught to new generations of Americans.

Both myths and cultures sprout from Americans searching for what they remember as a less complicated, better time when we had stronger families, stronger communities, stronger values. [1-1] We equate that time with the land and rural communities. And we create myths to make that linkage fit, because we must ignore horrible experiences of racial bigotry involving Blacks, Indians and new immigrants, and overlook the violence of Dodge City, Deadwood and Tombstone, to name but three wild frontier towns. In the agrarian myth of rural righteousness, we totally overlook the community leaders, the patriarchs and matriarchs of ethnic communities, school officials and theologians whose work and devotion built and sustained viable communities and neighborhoods within metropolitan areas over the past 200 years. Most outrageous, perhaps, is the idea we can wrap all rural people in a common mantle; I fail to see what the small–herd dairy farmer in Vermont or Wisconsin has in common with the Louisiana or Hawaiian sugar plantation owner, or with the California fruits and nuts grower who presides over family farm holdings from offices 30 stories up in Los Angeles.

I suspect something gets lost in the translation as rural–based literature is read in the city. It is a geographical warp that defies common sense. Literary reviews that heaped praise on Jane Smiley's Pulitzer Prize–winning novel, A Thousand Acres, saw the beauty and agony of the noble Iowa peasants and the book's Shakespearean theme. But did many urban reviewers or their readers recognize the Iowa family was making decisions and posturing for farm assets that would be valued at between $2 million and $3 million? [1-2]

South Dakota poet Linda Hasselstrom, though living in Wyoming in the mid–1990s, and South Dakota author, rancher and lay preacher Kathleen Norris both write about the lives, problems, beauty and the spiritual mind of rural people who, coincidentally, live on million–dollar cattle spreads. [1-3]

In Dakota: A Spiritual Geography, the rancher–writer Norris devotes extensive comment on the greed and callousness of rural Americans of the 1980s who watched their economy crumble. Well–financed ranch owners, like corporate executives in urban America, were waiting to pick up the assets of failed neighbors. [1–4] That book, a literary success in most parts of America, should dispel popular myths about who owns and operates modern agriculture. But it hasn't. Nor has The Last Farmer: An American Memoir, a delightful book by Washington, D.C., journalist Howard Kohn who, like his siblings, didn't want to take over the small family farm in Michigan. [1–5]

The Kohn farm was at least small; but unlike most rural farms, it was also sitting on oil reserves and was being squeezed by urban development.

There is great beauty to the land and to its people, and the authors reveal that, too. But despite the honesty of their writings, showing both beauty and blemishes, a myth about noble farm and ranch people – all hardworking, good family people – persists. I'm not sure if they are supposed to look like Ma and Pa Kettle, who lost the farm and stopped making movies in the 1950s, or the farm couple portrayed in Grant Wood's "American Gothic," who apparently retired from farming before machines started baling hay.

The reality today is that commercial–size family farms are substantial business investments. Most people living on former family farmsteads aren't commercial farmers, and they are both rich and poor, choosing to live where they do because rural housing is either inexpensive or conveniently distant from big cities. As the comedienne Dame Edna cautioned us in her "Dame Edna Christmas Experience," rebroadcast by National Public Television stations in 1993, "We mustn't judge Australia by the Australians." Here, she might need to reverse the quip: We mustn't judge rural Americans by rural America. But we do. Native Americans have long known, and noble peasants are discovering, that people from the countryside all look alike to people who have rural America trapped in an agrarian myth.

Politicians and public policy advocates keep the myths alive. They talk about the sacred mission of small, humble people, eking out an existence on the land while feeding the world. The long–running, 1986–1993 international trade negotiations under the General Agreement on Tariffs and Trade (GATT) showed these agrarian myths are not exclusively American. Europe has them. So does Japan. European Community Vice President Andreissen spoke of agrarian myths and the lingering public support they engender at a journalists' conference in Copenhagen. Strengthening the European myths, he said, are memories of food shortages during and after World War II. For older people in Europe, public policies dealing with farm surpluses are a welcome luxury brought by economic recovery in developed, or industrialized nations. [1–6]

The same is true in Japan, say Japanese visitors to the states, even though past public policies aimed at protecting that country's rice farmers made food unnecessarily expensive for most Japanese consumers. While older Japanese citizens remember food lines and shortages, younger members of the Diet (parliament) have resisted food trade liberalization because they remember more recent political interference in food supply lines. International trade and agribusiness consultant John Freivalds, in Grain Trade: The Key to World Power and Human Survival, gives a wonderful account of Japanese reactions to American soybean trade embargoes in the mid–1970s. [1–7] Among other reactive measures, Japan invested in South American agricultural development, most notably in Brazil and Argentina, speeding the inevitable development of South America's agricultural resources.

The GATT attempts to restore confidence in trading rules and in trading partners as reliable suppliers. But such confidence doesn't come easy – not with goods as basic to life as food. John Ochs, a political and press aide to former Agriculture Secretary John Block, recalls a casual encounter between the secretary and the Saudi Arabian agriculture minister while both were attending a United Nations Food and Agriculture Organization meeting in Rome. "Why are you spending so much money irrigating the desert to grow the world's most expensive wheat? We can sell it to you so much cheaper," the American secretary said to his Saudi counterpart. "Why are you spending one hundred dollars a barrel or more trying to get oil out of shale rock in Colorado?" responded the Saudi. "We will sell you cheaper oil." [1–8]

Trust among nations doesn't come easy. It is easier for societies to put their trust in noble peasants. We will keep our myths about an agrarian culture a while longer. Well into the Twenty–First Century, to be sure.

Perhaps forever. By looking at the myths, we can see why rural America needs to change its economic structure to survive. The culture of the Great Plains, and the Northwest states in particular, sheds light and offers models on how rural America might do it.

ಔಔಔ

Agriculture and its supposed way of life keep dying. Its death has little to do with food production, per se; rather, it visits us with changing economies of scale, new technologies and markets. Food production continues on.

Access to land, social standing, ways of life and agrarian politics are laid to rest. It has been that way since the Powhatan Indians showed

Jamestown settlers how to grow tobacco, then lost their land to the English colonists who wanted to expand their farms to serve their trans–Atlantic export markets.

Political leaders from Thomas Jefferson and Andrew Jackson on down to the present time warn of farming's imminent death, thus justifying the need for public policies or trade agreements.

Few chroniclers of past deaths to agriculture were as convincing as Richard Hofstadter. In his Pulitzer Prize–winning history, The Age of Reform, he saw agriculture passing from one era into another, connected with industrial development and a massive rural–to–urban population migration.

Professor Hofstadter also noted, humorously, how agrarian myths held fast even when economics and demographics changed. "Oddly enough, the agrarian myth came to be believed more widely and tenaciously as it became more fictional," he wrote in his 1955 classic. [1–9]

He recounted how, during Andrew Jackson's time, Pennsylvania Governor Joseph Ritner had campaign materials depicting him in broadcloth pants, silk vest and beaver hat while standing behind a farm plow. And later, he documented how the urbane Calvin Coolidge put a pair of overalls over his white shirt and pants for campaign photographers. While pretending he was haying, the presidential Pierce Arrow automobile and Secret Service agents were visible behind the hay rig. [1–10]

Such ridiculous scenes were revisited nearly every four years until recent elections. Presidents George Bush and Bill Clinton relied on jogging for exercise; to their credit, neither went looking for a cow pasture to run in for benefit of cameras. Their reasons for stopping such pandering, however, should trouble rural Americans. Politicians have discovered that only two percent of the U.S. population grows and raises food for the larger population. Just as agriculture, as we have known it, dies whenever someone invents a new technology that requires less farm labor or promotes greater productivity, it dies when candidates for national office hire campaign statisticians who run demographic data through computers.

It certainly died in Iowa in 1992. Senator Tom Harkin was a "native son" candidate for the Democratic presidential nomination so other candidates for his party's nomination stayed away. And President Bush's standing with Iowa Republicans wasn't threatened, so there was no big Republican campaign for farm votes there, either.

Iowa precinct caucuses are among the earliest contests for delegates in the presidential nomination process. Without Iowa, agriculture never

27

got much attention in the national campaign, complains Tom Cochrane, executive director of the Minnesota Agri–Growth Council.

That's the way it will be in future elections, too, warns Tom Sand, a political and press aide to former Agriculture Secretary Bob Bergland. He offers an indelicate but graphic political science assessment of why. The dwindling farm vote, he notes, is about evenly divided nationwide between Republicans and Democrats, leaving less than 20 percent of farm voters listed as independent, or "swing" voters. "Twenty percent of two percent (of voters) isn't very much,'' Mr. Sand said. "The gay and lesbian vote is bigger than the farm vote. The gay vote can be had." [1–11] Little wonder, then, that successful candidates for public office in 1992 chose to devote more time to civil rights issues in housing and the work place than in advocating higher farm loan rates or complex agricultural trade issues.

Another significant blow to agriculture, as we had known it, came in 1972 when the industrialized countries ended the Bretton Woods monetary agreement. Though most Americans didn't know of its existence, it established international financial institutions, such as the International Monetary Fund, and it had fixed the value of hard currencies to the value of gold after World War II. The coincidental result of abandoning the latter currency agreement was the biggest devaluation of the U.S. dollar since the Civil War, and the world came running to buy America's inexpensive grains and other farm commodities after converting their stronger currencies or gold into dollars. [1–12] "You grow it, I'll sell it," proclaimed Agriculture Secretary Earl Butz [1–13], who misread explosive U.S. farm exports as a permanent global supply–demand imbalance instead of a temporary response to monetary policies. [1–14] And he was an economist.

But the secretary was right in that a new era for American agriculture had been born. Foreign customers would keep coming for American grains and foodstuffs, provided that U.S. currency exchange rates made American exports affordable on world markets. This new era would not assure prosperity for farmers, however, and it ushered in an extremely painful period for agriculture. A land rush – more like a stampede – followed as farmers bid up the price of land to acquire more fields to grow exportable crops. With this land rush came the greatest inflationary period for American agriculture of the past century, culminating in the farm financial crash that ran from 1982 through 1987.

Both the buildup and the depression period that followed drove even more people from the land. Before the inflation cycle gave way to deflation, the American Agriculture Movement was born, reviving the William Jennings Bryan populism that Professor Hofstadter examined in his study of the period between 1890 and the Second World War. The AAM was a child of the frustration and tragedy of the 1970s inflation. Commodity prices didn't keep inflating with farmland values and operating costs. Young farmers, and especially young farmers just back from military duty in Vietnam, were victimized by paying inflated land prices to enter farming when farm income was stabilizing.

The AAM announced its arrival on the national scene by sponsoring "tractorcade" demonstrations in Washington during the late 1970s, calling on Congress and President Jimmy Carter to grant annual increases in government farm price support programs to cover the farmers' ever-increasing expenses. Earl Butz was long gone from the Agriculture Department, but large numbers of farmers still followed his gospel: "Buy farmland. They ain't making any more of it." [1–15]

Farmers did; land costs soared. The "tractorcades" of the late 1970s had everything but a "Cross of Gold" speech from a latter day Senator Bryan. Though nobody said it, it was a rebirth of the pro–inflation movement of the earlier era [1–16].

There were charming moments to remember from those farm demonstrations. When a serious snowstorm swept in on Washington one January, the farmers used their tractors to clear streets of snow – a public service the city wasn't prepared to do. The farmers became heroes to Washington residents, even to those who were confused about why the farmers were in town.

Bill Vance, an editor at Knight-Ridder Newspapers, and I witnessed this bewilderment when working one snowy night after President Jimmy Carter delivered a State of the Union message. A streetwalker stopped two AAM farmers on the sidewalk near the National Press Building to ask, "What do you boys want?" The farmers looked at each other, then buckled over in laughter.

"We want parity!" shouted one farmer. "Haven't you heard?"

The young lady from Washington, wading through deep, fresh snow in knee–high boots and hotpants, didn't have a clue about what the farmers said or why they were laughing.

She walked away. But then, I've seen people of higher stature in Congress and the White House do the same thing at the mere mention of loan rates, target prices and parity – that esoteric economic measurement of farm purchasing power based on the years 1910 to 1914 that were considered the golden age for American agriculture.

It's not a new phenomenon. Everyone has a good word for farmers, even when they don't want to be burdened with understanding farm issues. Economist and public policy expert C. Ford Runge, for instance, tells of the time President John F. Kennedy called his Agriculture secretary, Orville Freeman, to a White House meeting with other cabinet officers. "I don't want to hear about farm policy from anyone other than you," the president reassured his secretary. "Come to think of it, Orville, I don't even want to hear a lot about it from you."

An even more humorous display of presidential uneasiness, if not indifference, to American farm policy came in 1984, in the midst of that decade's farm financial crisis. President Ronald Reagan was running for re-election that year. His Agriculture secretary, John Block, prevailed on White House aides to schedule a meeting with the President for visiting agricultural journalists. The result was one of the greatest, hour-long comedy routines ever performed at the White House.

The President said he was sympathetic with the "farm problem" because he, too, owned a cattle ranch in California. But it was hard to gauge the sympathy. He told jokes about all the programs Secretary Block had over at the Agriculture Department, claiming there was a program for every farmer.

Some of those programs had helped him out as well, he said. For instance, President Reagan recalled a time when he was Governor of California when federal animal health inspectors called and said they needed to examine his cattle.

A disease was threatening America's beef livestock herd, leaving breeding stock unable to reproduce. Not grasping the contagious nature of the disease, the President laughed at the inspection. "I only had steers," he said.

The animals were found to be healthy, the President recalled. "That was a relief, until I asked what would happen if they were sick." Federal inspectors informed him a disease indemnity program would pay him $500 for each infected animal, though he could sell them on the livestock market because the disease didn't cause a problem for the meat. "They asked me, 'Have you any other questions, Governor?' So I asked them, 'Where can I buy me some sick steers?'"

I was sitting with Audrey Mackiewicz, the long-time secretary and treasurer of the National Association of Agricultural Journalists from Sandusky, Ohio. Across the table from us were President Reagan and Secretary Block. We debated later which was funnier, the President's sick steer joke or how sick Secretary Block looked after the President made light of his department's disease eradication program.

It got better. The President was only warming up. He told about how he and Nancy Reagan moved to the ranch after serving as governor, and how they decided to become "farmers" as well as "ranchers." "We decided we should get some chickens so we could have real fresh eggs ... those eggs where the yolks sit right up on the whites," he said. That meant getting an architect to design a chicken house, a contractor to build it, and building and zoning permits to keep chickens on his urban ranch estate.

Then he had to buy chickens. That meant bringing back the animal health inspectors to make sure the chickens were healthy and their eggs fit to eat. "The next April, our tax accountant asked how we liked the eggs," President Reagan said. "I said they were wonderful. The accountant said that was good because they were costing me $80 an egg."

This was about the same time Texas Agriculture Commissioner Jim Hightower, a humorist from the opposite political aisle, was touring the country and telling farm audiences, "The problem with President Reagan is, he thinks a good farm program is 'Hee–Haw.'" There may have been some truth to that, but it didn't matter. President Reagan won re–election in a landslide. The farm crisis of the mid–1980s had no impact on the electorate. As Mr. Sand said, 20 percent of two percent isn't much. The AAM still goes to Washington to lobby, and it still wants parity. Its members call for government price guarantees, but not inflation, to achieve it. That doesn't makes the AAM substantially different from other broad–based farm organizations.

At the risk of offending all farm groups, I would say the groups' main difference in appeals for federal assistance is in selecting which door to use when receiving subsidies. The AAM, National Farmers Union and, to a lesser extent, the National Farmers Organization, are willing to take their subsidies through the front door, convinced that a strong farm economy is in the national interest. Thus, they argue, orderly marketing and income stabilizing programs are also in America's interest. The American Farm Bureau Federation isn't as comfortable with direct government assistance so it prefers its subsidies through the back door, such as tax breaks or low–cost access to public lands. This has made the Farm Bureau one of the largest member organizations in the nation's tax avoidance movement in informal partnership with large industrial trade associations and professional groups. When large farmers and large manufacturers argue for "a level playing field" to compete in international trade, they don't discuss how America's comparatively low income and property taxes may be distorting trade. [1–17]

Reality says the farm organizations won't win many future skirmishes over public policies. The Uruguay Round of GATT trade negotiations, completed late in 1993, dictates less, not more, public subsidies for agriculture. Federal budget deficits and pressure from the general public dictate there won't be favorable tax breaks coming for either large land holders or large manufacturers. Moreover, the lawmakers who deal with agrarian policies are armed with demographic and economic data assembled by the Agriculture Department's Economic Research Service. They, too, can count and see that farm income programs favor large farmers, do little to assist the rural poor and small farmers, and that there may be an inconsistency in supporting farm income for people who are already complaining about their tax burden under the graduated income tax.

In early January, 1994, the authoritative AgWeek farm magazine at Grand Forks, North Dakota, raised the curtain on public policy debates for the 1995 farm bill. In proclaiming that farm programs are indeed at "a crossroads," [1-18] AgWeek said the GATT agreement assures that all 117 participating countries will begin dismantling farm subsidies and trade barriers, "It's just about given that water quality and other environmental concerns will get top billing in the 1995 farm legislation," the writers added. And, they concluded, "You can read into this trend at least one fact about future farming: Producers will be forced to change how they grow the country's leading crops. Producers – both in and out of the farm program – will have to be increasingly selective about the varieties they choose and about where and when they plant them. The freedom to do anything one wanted is going to be curbed." [1-19]

I would argue that technological changes in the 1990s are imposing even greater restrictions on how one farms and how one uses the new generation of genetics, nutrients and chemicals. Livestock producers are being told how and when to market their animals, and where they can locate barns and feedlots. They are also being told what the animals should look like, and what their weight should be when brought to market. The noble peasant will be praised for making such changes and patted on the back while changing. But as the noble savage learned long ago, you can't eat or bank the nice things people say about you. All the while these changes are occurring, farmers will lose an element of individual control. The agrarian myth won't go away. Society so loves the myth it won't stop trying to take care of the noble peasant. But the focus is shifting. Even as income support programs are reduced, it will attempt to make farmers a ward of the state. It wants to protect them from themselves.

ഇംഇംഇം

"Western agriculture is critical to our food supply and to consumers in both industrialized and developing nations. Yet many Western farmers, like those elsewhere in this country, have suffered through years of economic hardship, which – if

32

unchecked – could threaten our family farm system. We cannot ignore consumer interests, and America's vital role in the global food economy, but we must never forget our overriding interest in the long term health and stability of our agricultural system. That system must be preserved."

Vice President Walter Mondale, From speech delivered Jan. 12, 1978, at Washington State University, Pullman, Washington.

While myths are built on stereotypes and can be applied with a broad brush across the landscape, cultures are built by people and can be unique to a region, a community, even a company. The historic Northwest region of the United States, which encompasses everywhere north and west of Chicago, has a distinctly different culture from other regions of the country. It comes from being primarily rural despite having urban areas around Milwaukee, Des Moines, the Twin Cities, Seattle and Portland. When Nebraska is considered part of the region, Omaha is another metropolitan center. While the region's people embrace the ideas handed down to all rural Americans by the Jeffersons and Jacksons, the Northwest has a more narrowly defined culture that was shaped by immigrants, period of settlement and development, distance from markets and shared experiences on the frontier. It is a culture that found cooperatives to be a useful tool for development and problem solving in earlier years. This culture can be shared with other regions of the country. But for it to be a useful tool in the future, it must be understood and removed from the agrarian myth.

ഇ൭ഇ൭ഇ൭

The Northwest's regional differences became apparent in 1978 when Vice President Walter Mondale led a rescue mission to the West in an attempt to restore political relations between the Western states and the White House. Political pundits had prematurely proclaimed the end of regional politics two years earlier when Georgia Governor Jimmy Carter Jimmy, was elected president. Neither President Gerald Ford nor President Carter used campaign strategies to arouse racial fears or regional animosities. This gave rise to a short–lived theory that America had become a nation of one people, North, South, East and West. That was before youthful White House aides, working with Eastern–based environmentalists, prepared a "hit list" of water projects that were to be stricken from government public works programs.

Most of these public works projects would die later from lack of budget support when California Governor Ronald Reagan became

33

became president. No matter. The "hit list" brought the wrath of the West down on President Carter. It was a political disaster of Grand Canyon proportions.

Agriculture Secretary Bob Bergland, who could speak the language of farmers, and Interior Secretary Cecil Andrus, a former Idaho governor who could speak the language of foresters and environmentalists, accompanied the vice president on a swing through Western states. It was a losing battle apparent to all by the time the White House party passed through Montana, Wyoming, Colorado, Utah and Idaho and reached Pullman, Washington, where Vice President Mondale gave the only major address of the tour. The vice president responded logically. He gave a classic version of the all-weather political speech – so complete that I saved a copy. It had nearly every platitude you could drop on a Western audience. It offered no promise of a change of policies by the federal government. It was simply that time–tested political message: Your government loves you and is here to help you.

To his credit, the Vice President did a masterful job of humoring and charming audiences at informal gatherings and town hall meetings. But the Westerners would not be swayed. The damage had been done and was beyond repair. The trip, however, showed that America hasn't become a unified, single country. There are distinct regions with profoundly different political and public policy agendas.

Sixty years ago, a frustrated Connecticut senator described the Senate's political mavericks from Nebraska, Wisconsin and points westward as "the Sons of the Wild Jackass." It might still be a fitting description. My home state, Minnesota, has hybrid forms of both Republicans and Democrats – a source of irritation for some resident ideologues of the two national parties. We have Democratic–Farmer–Laborites and Independent–Republicans. More often than not, our politics track more closely with political divisions found in Canada and Northern Europe than in other regions of the country. From personal observation, but not serious study, I find that the factions of Independent–Republicans capable of electing statewide office holders have a lot in common with the Conservatives, or Tories, in Canada. There are just enough Reagan–style American conservatives in the state party, however, to mess up Republican election chances with intra–party fights. And the DFL party that gave rise to the Humphreys, Mondales, McCarthys, Freemans and Berglands bears striking resemblance to the not–so–conservative Conservative parties and national Social Democrat parties in Northern Europe. It currently has factions caucusing at local conventions that defy easy identification with other political movements.

Leading political scientists such as V.O. Key and Hans J. Morgenthau always insisted that America's 50 states produce 100 political laboratories that are state parties. It's true in the Northwest states, and the laboratories even produce hybrids. Allen I. Olson, for instance, remembers being visited by a Wall Street Journal writer after he was elected governor of North Dakota in 1978. After several years of Democratic–Nonpartisan League rule, the Republicans gained a majority of executive positions in state government. Now that the "Socialists" were out of power, the reporter wanted to know, would North Dakota sell off its state–owned Bank of North Dakota, the North Dakota mill and elevator and other state–owned properties? "I think we offended the Journal," the former governor recalls. "We asked why should we do anything so radical as that." [1–20]

Northwest residents inherited political approaches to problem solving from immigrant ancestors who settled on the frontier, and from Native Americans who were already there.

They turn to government, at one level, when they need help to develop resources and infrastructure they can't develop themselves. They pool resources at the community level to do things when cooperative, collective action and capital is needed and possible. They share a culture that allows them to be both liberal and conservative, usually at the same time.

ഇൽഇൽഇൽ

It is time to revive this shared cultural heritage to turn the geopolitical differences of the Northwest region into a strength for future rural economic development. It is a culture that would work for other regions of the country as well. But I have traveled often enough in the Southeast seaboard states, the South and Southwest to know regional approaches to problem solving differ greatly. Moreover, while I believe the Deep South states have economic and geographic advantages for building their rural communities and economy, I see that region as the exact opposite of the Northwest: The South must learn to forget its past. [1–21] It has the infrastructure in place to support people living on the land and in rural communities, if it can learn to work cooperatively within those communities. The rural Northwest, on the other hand, must constantly build new types of infrastructure, and new business systems, so commodities and value–added manufactured products may reach distant domestic and export markets at competitive prices. To do so, it must recall what worked on the frontier.

Such efforts at making the frontier work has created a culture closely tied to trade. And it is important to note that it didn't start with white settlement. In fact, it predates American independence. Chief Wabasha II, the second in a dynasty of great chiefs from a settlement along the Mississippi River, led a trade mission across the Great Lakes to Montreal in 1774 to petition the French governor–general to keep trade routes open to the West. Trade was being interrupted by French–English wars in the East that embroiled the Indian tribes as well.

Helen Willard, a veteran Washington state journalist at Prosser in the Yakima Valley, and an author whose writings have chronicled developments with area tribes, says the Pacific Northwest tribes had trade relations in what are now Washington, Oregon and Canada's British Columbia long before American independence. [1–22] Even in the vast prairie lands of the Dakota, or Sioux people, lived nonmigratory tribes of farmers and craftsmen, notably the Mandan and Pawnee tribes, that needed avenues of commerce to reach a larger world. [1–23]

White settlement in the Northwest reshaped the political geography. It didn't change the region's need for trade relations far from home. Instead, it added new influences into the developing Northwest culture.

Given the origins of most new arrivals, experiments with territorial, state and local cooperation drew heavily from literature and first–hand experiences brought over from Northern Europe by immigrants. Waves of immigrants swept into the Northwest from about the time of the American Civil War to shortly before the First World War. Railroads, land companies and local governments promoted the migration, promising free and inexpensive land in an unspoiled "Eden" on virgin prairies. Indeed, the propaganda was such nonsense that historian Hiram M. Drache has pondered which government action inflicted the greater pain and suffering on American citizens: the Civil War or the Homestead Act. [1–24] Later migrations followed the spread of business and railroads westward from Wisconsin and Minnesota, and back inland from Portland and Seattle. The intra–regional movements of people were so great that a Washington state demographer, analyzing the 1970 census for his state, still listed "Minnesotans" as the third largest ethnic group in his state. [1–25] Those population movements not only built a shared culture for the region, they established commercial ties and trade patterns that remain in place today. [1–26]

ಬಿಬಿಬಿ

In 1994, American cooperative businesses sent delegations to Great Britain to celebrate the 150th anniversary of the founding of Rochdale

Equitable Pioneers Society. Rochdale, near Manchester in Northern England, is regarded as the first modern era cooperative business. [1–27] But it, like institutions started in the American Northwest, can trace its origins to a period of time when farmers, workers and their political leaders were reacting to economic stress brought by the Industrial Revolution. Professor Brett Fairbairn at the Centre for the Study of Cooperatives at the University of Saskatchewan, notes that Rochdale wasn't a farm co–op. "Its origins were not agrarian at all, but based in the working class, trade union, suffrage and factory reform movements. The Industrial Revolution was also having an impact on agriculture, and agricultural reform and cooperation were responses to social and economic change in the same way Rochdale was." [1–28]

In an exchange of notes during summer of 1994, Professor Fairbairn drew an interesting parallel between the time of Rochdale's founding and now. "You could say that all cooperatives have their origins as responses by people to economic restructuring," he wrote. "I've argued elsewhere that the free trade liberalism of the 1860s, which helped spread industrial change, made up a first round of what we now call 'globalization'." From that, he said, came the world's first rural credit cooperative in Germany (1862), the world's first rural consumers association in Denmark (1866) and various other responses to "the beginning of globalization in the 1860s."

In the countryside, these responses were tied to agrarian reform movements that spread throughout Northern Europe. Reform church leaders often gave strength to these movements. New national constitutions with strong agrarian protections were adopted in Denmark and the Netherlands.

Feudalism was ending with land reform policies stripping crowns and churches of massive land holdings. Manorialism, a variant of feudalism, is still found in parts of the United Kingdom, and surviving forms of it continue in what were British colonies – including the plantations of the "landed gentry" that control farmland production in areas of the American South. Forms of "manorialism" also dominate agricultural production in parts of Latin America and Africa, as well. [1–29] Land reform rebellions, and formation of producer–controlled business enterprises, were part of the political and economic transformation occurring in Northern Europe shortly before the American Civil War. For many Europeans, notably the Scandinavians, Germans, Irish and Dutch, immigration offered a faster route to realizing the peasant dreams we've come to claim as the "American dream."[1–30]

The immigrants inspired communities to pool resources for doing things the government and private investors didn't do on the Northwest frontier. Cooperatives were formed, ranging from dairy and grain marketing businesses to community cemeteries, parks, restaurants, libraries, water systems, newspapers, and, eventually, utilities such as electricity and telephone co–ops. And they created the "frontier" as defined by University of Wisconsin historian Frederick Jackson Turner, who saw a democratic egalitarianism rising from the frontier as well as a "safety valve" that allowed Europe's and America's surplus populations to migrate to new opportunities in new hinterlands. [1-31]

Professor Turner's theories are challenged by today's historians and social scientists, particularly his notions about frontier egalitarianism. [1-32] But his general theme of the frontier serving as a safety valve is being given new currency by scholars, both for its opportunities for migrations around the world and for providing trade outlets for generating wealth from surplus productivity. [1-33].

If one accepts the notion that the sparsely populated but surplus producing Northwest states are now fully developed, more attention must be given to Professor Turner's "safety valve" theory. The Northwest must look to developing, Third World countries to absorb its commodities, manufactured goods and services. [See Chapter 6]

But that isn't enough. Inherent in accepting the Northwest as fully developed is an admission of failure – a failure of not finding productive ways to keep people living on the land and in financially healthy, viable rural towns. The same can be said of rural communities in other parts of the country as well, and most certainly it is true of the Western Prairie Provinces of Canada. Agricultural science no longer requires large numbers of people to produce bulk farm commodities for shipping to processors and animal feeders in other regions or countries. Ways must be found to utilize those raw material resources more closely to where they are produced, raising their economic value in their nearby community and providing jobs for people living in towns and on rural, former farmsteads.

ഇന്ദ്രഇന്ദ്രഇന്ദ്ര

Models for doing so may not be easily found in experiences of other regions of the country that are closer to large domestic population centers. But models do exist if today's rural leaders look abroad as their ancestors did earlier. They can look to New Zealand, Australia, north to Canada, and any number of European countries that have both developed industrial and agricultural sectors.

Oregon doesn't have the huge flocks of sheep, but it does have other similarities to New Zealand. Wisconsin is a lot like Austria, or parts of Germany, without the mountains. Minnesota is incredibly similar to Denmark, in every way except geography. This is especially apparent when looking at Denmark's and Minnesota's agricultural, industrial and technological sectors, and their political environment [1–34]. Washington state? It's as if a large group of Norwegians packed up and moved to another area of the world where they could also have beautiful mountains, forests, rich plains to grow food, and natural harbors and fertile fisheries to work the seas. The presence of Boeing, the state's dominant industry, distorts modern day comparisons to some degree. But so do the successful industrial companies back in Oslo.

Montana and the Dakotas are harder to place with partner cultures and resources abroad, given their small populations and climatic limitations that include short growing seasons and shortages of rainfall. But they have "foreign" models to choose from, right across their northern border in the Prairie Provinces of Canada, notes Professor Fairbairn. They share the same geography influences, the same historical development, the same crops and – to an increasing extent – the same markets, and in many cases they are brothers and cousins from the same generation of immigrants. In addition, their social and political culture is unmistakably transplanted from Northern Europe.

Piecemeal, there are models for rural and agrarian development that may be consistent with our own regional and cultural heritage. Dan Looker, business editor of Successful Farming, pointed to potential models in an April, 1993, editorial in his highly regarded farm magazine:

ဆာဆာဆာ

"As the turn of the century nears, smaller farms have a tougher time supporting a family on the thin margins they get from raising grain. There is a glut in the world market, with grain coming from nations that use cheap labor on huge farms, and from developing countries that are building new transportation to ship crops.

"A description of the Midwest? No, it's Denmark a century ago, when exporting wheat and barley was starting to look like a dumb way to make a living. The huge farms of their competitors weren't in Brazil, but in Russia, where peasants reaped a bounty of wheat. And there was that brash new industrial power – not Taiwan, not Thailand, but the United States – with new railroads, irrigation in the West, and millions of enthusiastic, young farm families.

"So the Danes began to feed more grain to hogs, discovering the importance of 'value–added' products long before our agricultural leaders coined that buzzword.

"Today, Denmark exports 80 percent of its pork, with a value of $3 billion to $4 billion that represents a tenth of that nation's exports. And farmers control all of that through five co–ops that coordinate breeding and feeding programs and slaughter and process pork. 'From farm to table, the farmer is in control,' says Erik Klindt Andersen, agricultural counselor for the Royal Danish Embassy in Washington, D.C. The farmers own the production facilities, the hogs and the slaughter plants. It's a form of vertical integration from the ground up. It has allowed Danish farmers to receive premiums for hogs that average more than 59 percent lean. And that tough standard has helped the Danes become the world's leading exporter of pork." [1–35]

ഇൽഇൽ

The circle is unbroken. We are back to studying the Danes who have contributed to our Midwest and Northwest culture in far greater proportion than their number of immigrants would suggest. They can show us again how to cope with changing world markets and technology.

It is ironic, perhaps, that socially–conscious people in Midwestern farm groups who have borrowed so heavily from Danish and Scandinavian social theory have distanced themselves from progressive steps being taken in Europe. These farmers, especially those aligned with National Farmers Union state organizations, have taken a conservative view of the structure of agriculture in states such as Minnesota and have opposed cooperative livestock developments. They are at the forefront of change, however, next door in North Dakota. Where they are opponents of cooperative production schemes, they see these new developments as an affront to individual farm ownership and management – not as a progressive adjustment to changing markets and farm economics.

They have a point, until one looks at the alternatives – larger and fewer grain farms while livestock and poultry production is moved from the Midwest and consolidated under agribusiness firms in North Carolina, Missouri and Oklahoma. [1–36] These farmers have allies in opposing cooperative production, including people who don't want livestock production near their towns, environmentalists who fear large concentrations of animals on a given site, and others whose agenda is more diverse or harder to read. But among the latter are well–intentioned advocates of a family farm image who want to perpetuate a peasant society that has disappeared from the landscape, but not from the agrarian myth.

How has sophisticated, cooperative hog production changed Denmark? Mr. Looker notes that the small Scandinavian country, with a population approximately equal to Minnesota's and about half the land, has managed to sustain 100,000 farm units. That was what Minnesota had before the start of the 1980s farm depression, during which about 20,000 farm units went out of business. Will a Danish model work in American states? Yes, in some states. But opponents of new technologies and business structures – while trying to preserve the family farm – will discourage cooperative ventures and thus contribute to the exodus from the land. In early 1994, for instance, the corn farmers who own Minnesota Corn Processors Inc. cooperative at Marshall, Minnesota, were planning to build a large beef cattle feedlot across the border in South Dakota. The animals will be fed the gluten byproducts from their corn wetmilling plant. But the farmers didn't want the political fight that was certain to be waged if they tried to build the feedlot in their own communities.

Mr. Looker, in his Successful Farming editorial, correctly noted that some U.S. states don't have a cooperative tradition. What's more, he quoted Paul Lasley, the Iowa State University rural sociologist as saying the spirit of cooperation is being lost. "In the past 50 years, farming has become much more of an individualized activity." [1–37]. Meanwhile, Mr. Looker said meat packing and food processing companies are taking over more of the meat production industry in states where there are no barriers to farm production ownership. This trend, too, will limit farm management alternatives in future years.

ΒΟΒΟΒΟ

There is an agrarian culture in the Northwest states that can be useful for communities wanting to chart out a course for survival in the Twenty–First Century. I worry, however, that we may be losing our shared, Northwestern culture and our ability to work within it to think like pioneers and entrepreneurs. I said it once before:

"I grew up in a rough neighborhood. Droughts, hail, tornadoes, grasshoppers, blights and icy roads were facts of life.

"We were poor by national comparisons, but we had a healthy naivete so we didn't know it. When you don't know your place in society, you can't be held back by it.

"I worry now that rural Minnesota may be losing its naivete, accepting a "place," or role.

The economic hardship that has fallen on farms and rural communities may be lowering expectations and limiting self identities. It shouldn't happen, but it probably is in certain communities.

"This personal concern began a couple of years ago when Mikhail Gorbachev emerged as Soviet general secretary. Major national publications scoured the academic and corporate landscape trying to find those few people who knew this surprise choice of the Soviet power structure.

"It was pleasing to see the Grand Forks Herald, (N.D.) not have that problem as it went out into the Red River Valley and found wheat growers and elevator operators who knew Gorbachev personally from his days as agriculture minister. But in subsequent generations and future changes of leadership here and around the world, will rural Minnesotans know the players or only what they read about them in newspapers?

"This concern is troubling me again as thoughts turn back to childhood during the holiday season and as I read my hometown newspaper, the Kerkhoven Banner.

"A recent edition of the Banner tells about my high school friend, Lee Roisum, who grew up on a farm at nearby Sunburg and is now a baritone for the New York Grand Opera.

"A stagebill published by New York's Lincoln Center reported on the All Verdi Gala at Alice Tully Hall, featuring Enrio Di Giuseppe from the Metropolitan Opera and Roisum from Sunburg.

"The stagebill noted that Roisum was about to sing a premier with Gabraella Tucci from Italy's La Scalla, and that he had recently performed such roles as Tonio in I Pagliacci, Enrico in Lucia di Lammermoor, the elder Germont in La Traviata, Amonasaro in Aida and had repeated his favorite role, Rigoletto.

"I thought about Lee's great voice, too strong for him to sing in what is now the Kerkhoven–Murdock–Sunburg school choir. And how he knocked windows out of old Eastman Hall at St. Cloud State (University) when he visited me and sang at a charity fundraiser. And how he let a couple of barn swallows loose in a theater during a showing of Alfred Hitchcock's "The Birds."

"This is the Lee Roisum who always insisted on having fun. He knew what he wanted, and he wasn't concerned about the mathematical chances of a Sunburg farm boy growing up to sing opera in New York City. As an athlete, he once stood on second base, raised two fingers toward heaven, and shouted in full voice, "Hey, up there, we need two big ones!"

"The next batter delivered a bloop single over the third baseman's head. driving in the tying run from third. Roisum, running wildly on his own, scored from second to produce a rare small school district tournament victory over big Willmar.

"The coach, an Augsburg College graduate, thought it nice that Roisum had reaffirmed the power of prayer. But reverence and the hit–and–run were discussed in detail at the next team practice.

"There were so many others like him at Kerkhoven, Murdock and Sunburg, choosing to follow different callings into adulthood – hospital administrators, clergy, teachers, farmers, CPAs, veterinarians, doctors, nurses, business people.

"We lost friends and relatives in Vietnam, and some of our contemporaries have been badly hurt by the farm depression. But most have more than survived.

"There was Jerry Koosman a few towns away at Appleton, who thought he could pitch in the big leagues, so he did. And Peter Henrickson, who built a concert–size harpsichord in his parents' basement at Clarkfield because there weren't many of them in western Minnesota.

"Our generation out on the prairie produced an extraordinary assortment of published poets, essayists and other types of writers, and some great thinkers who have difficulty with words.

"There's the Perham farm boy, armed with a master's degree in philosophy, who has tried for 20 years to find words describing Scandinavian non–verbal communications.

"There's Bill Holm [1–38], now at Southwest State (University), who began writing poetry successfully when he ran out of books to read at the Minneota library.

"There's Jon Hassler [1–39] at St. John's University, Minnesota's best writer, who combines rural Minnesota Catholicism and Protestant existentialism every time he writes a book. It would take Soren Kierkegaard, if he were still around, or Gene McCarthy, if he felt inclined, to explain what burns in the hearts and moves the fertile minds of people such as Holm and Hassler.

"Perhaps it is Minnesota's Northern European–like climate and the Northern European origins of the vast majority of rural residents.

"Perhaps we have created a new Northern European culture here that combines local theological and personal philosophies. Perhaps we have trouble recognizing our own limits when the horizon is always so far off on the northern prairie.

"People don't want to die in western Minnesota, even if it means a better life without droughts, blizzards, corn borers or depressions. Still, they try to arrange in advance for Dave Frederickson, a farmer, educator and state senator from Murdock, to sing at their funerals.

"It's nice like this. It's productive. It's rewarding for all who notice how distant it is to the far horizon. There is beauty, and much music, and great words at work with farm sciences.

"Lee Roisum isn't Rigoletto. He is what we all still can be, and what future generations of rural Minnesotans should be free to become."[1–40]

Chapter 1 Footnotes

1. This is a reoccurring theme in modern literature, though not often supported by research. Substance for this belief, however, can be found in a number of books by Kentucky essayist Wendell Berry. Among the more enjoyable Berry offerings: Home Economics. North Point Press, Berkeley, Calif. 1987.

2. Smiley, Jane. A Thousand Acres. Fawcett Columbine, New York. 1991.

3. Norris, Kathleen. Dakota: A Spiritual Geography. Ticknor & Fields, New York. 1993; and Hasselstrom, Linda. Windbreak: A Woman Rancher on the Northern Plains. Barn Owl Books, Berkeley, Calif. 1987.

4. Norris, Dakota.

5. Kohn, Howard. The Last Farmer: An American Memoir. Harper & Row, New York. 1988.

6. Andriessen, Frans, vice president of the Commission of the European Communities, July 18, 1987; from an address and comments given to reporters' at Copenhagen, Denmark.

7. Freivalds, John. Grain Trade: The Key to World Power and Human Survival. Stein and Day, New York. 1976.

8. As retold by John Ochs, special assistant to the secretary, U.S. Department of Agriculture.

9. Hofstadter, Richard. The Age of Reform. Vintage Books (Random House), New York. 1955.

10. Ibid.

11. Sand, Thomas R., in comments from a discussion, Cedar Street Cafe, St. Paul, Minn., 1992.

12. A widely accepted conclusion among agricultural and trade economists, but not widely understood at the time of the post–Bretton Woods devaluation. Among the early advocates of the monetary policy–trade linkage was G. Edward Schuh, dean of the Hubert H. Humphrey Institute of public affairs, who served as an economist in the Ford and Carter administrations and formerly was director of the agriculture and rural development department at The World Bank.

13. Butz, Earl, from comments the Agriculture secretary made in 1974 to justify eliminating mandatory acreage set–aside programs.

14. Agriculture Secretary Butz wasn't alone in not understanding the temporary nature of America's explosive export trade. Most people associated with agriculture wanted to believe that the world would stand in line to buy American farm commodities. And hunger organizations and private research groups, notably the Worldwatch Institute in Washington, D.C., were convinced world population had crossed the "Malthusian Curve" and would no longer be able to feed itself.

15. He had a million of them, as they used to say in vaudeville days. This was one of his more widely published one–liners.

16. Perhaps unbeknownst to the speakers using the quote, one often repeated statement used at tractorcade protest demonstrations was taken directly from

William Jennings Bryan's "Cross of Gold" speech: "Burn down your cities and leave our farms, and your cities will spring up again as if by magic; but destroy our farms, and the grass will grow in the streets of every city in the country." (This wonderful example of pandering to an agrarian audience made no sense in Bryan's time, and is counter to all demographic and economic trends of the past century. – The author.)

17. Nothing risks infuriating Americans more that suggesting they have a light tax burden. But studies by the Union Bank of Switzerland in Geneva and the Organization for Economic Cooperation and Development (OECD) in Paris do show that America has both comparatively low taxes and low labor costs among leading industrialized nations.

18. AgWeek. Grand Forks, N.D., Jan. 3, 1994.

19. Ibid.

20. Olson, Allen I., from a 1993 interview, St. Paul, Minn.

21. Recent elections in the South, with the 1992 Louisiana gubernatorial contest offering a vivid example, suggest that race and station in life are still important, divisive influences in many states. The author fears this precludes the communities from effective cooperative investment and development activities.

22. Recalled in numerous histories of the state of Minnesota.

23. Yenne, Bill. The Encyclopedia of North American Indian Tribes. Arch Cape Press (Crown Publishers), 1986.

24. Drache, Hiram. Koochiching. The Interstate Printers & Publishers, Danville, Ill. 1984.

25. The author witnessed a discussion of 1970 Washington state census data among Vice President Walter Mondale and state officials, Jan. 12, 1978, at Pullman, Wash., while traveling with the White House party discussed in this chapter.

26. From state histories. A particularly helpful one is: Wilkins, Robert P. and Wynona H. North Dakota: A Bicentennial History. W.W. Norton & Co. New York, New York. 1977.

27. Thompson, David. "Rochdale Revealed." Cooperative Business Journal, October, 1993, Washington, D.C.

28. Fairbairn, Brett. Centre for the Study of Cooperatives. University of Saskatchewan, Saskatoon, Sask. From personal correspondence, August, 1994. And for a particularly thorough and concise history of the originis of cooperatives in Europe and North America, see: Fairbarin. "Cooperatives and Globalization: Market–driven Change and the Origins of Cooperatives in the Nineteenth and Twentieth Centuries." Printed in Globalization and Relevance of Cooperatives. Centre for the Study of Cooperatives. University of Saskatchewan, Saskatoon, Sask. 1991.

29. As defined in: Stavenhagen, Rodolfo. Social Classes in Agrarian Societies. Anchor Press/Doubleday, Garden City, N.Y. 1975; and Kopytoff, Igor (editor). The African Frontier: The Reproduction of Traditional African Societies. Indiana University Press, Bloomington, Indiana. 1989.

30. Among good examples, see: Amato, Joseph A. Servants of the Land. Crossings Press, Marshall, Minn. 1990; and Radzilowski, John. Out on the Wind. Crossings Press, Marshall, Minn. 1992.

31. Frederick Jackson Turner has caused volumes to be written about his theories of the role and culture of the frontier. The University of Wisconsin historian first published monographs on the subject in 1893. They were later published in books, in 1920 and 1947, which can be located in comprehensive academic and research libraries.

32. The author listened to an unlikely discussion of Turner's oversights one Sunday afternoon in 1991 during a volleyball game on a farm near Hudson, Wis.

Among the players were academics from the University of Minnesota and University of Wisconsin – River Falls.

33. Kopytoff. The African Frontier.

34. From personal observation and discussions with Danish government, academic and agriculture leaders in Denmark. This is also the observation of Robert Buckler, president of Issues Strategies Group, from his frequent visits to clients in Denmark.

35. Looker, Dan. "Business." Successful Farming. April, 1993.

36. Pig placements and slaughter reports, Economic Research Service, U.S. Department of Agriculture, 1991, 1992, 1993.

37. Looker, Successful Farming.

38. Among other books currently in print, Holm is author of Coming Home Crazy and The Dead Get By With Everything.

39. Among other books currently in print, Hassler is author of Staggerford and A Green Journey.

40. Egerstrom, Lee. "Minnesotans lose naivete by setting limits for themselves," St. Paul Pioneer Press. Dec. 21, 1987.

CHAPTER 2:
REVIVING THE "GOOD OLD DAYS"
AFTER EVERYTHING'S CHANGED

"He's an odd sort of idealist. He wants to be a bona fide traditionalist. To be a singer of the curing ceremonials.

"Not just a shaman, but to be a really effective one." Leaphorn paused, looking for some general statement to sum this up, and his attitude toward it. "It makes any sort of taboo more powerful than it would be to me – and probably to you. Officer Chee wants to save his people from the future."

Lieutenant Joe Leaphorn in Sacred Clowns (fiction), by Tony Hillerman. Harper–Collins, 1993.

All of rural America is well populated by shamans. They are heroes on the Navajo reservations in Tony Hillerman's beloved Southwest. They live and work among us in the Upper Midwest, the Northern Prairie states and Pacific Northwest as well. These medicine men and women can be found in rural Chambers of Commerce, church pulpits, farm organizations and agricultural trade and promotion councils. They're in forests and mines, and in the rural towns that support raising and harvesting of natural resources. They are motivated by a high calling to protect us from the future.

We need these people. When we remember who we are and where we came from, we also don't forget to care for one another and for our environment. We become neighbors. Good neighbors shape us into tribes that we call communities. And from these good neighbors and our tribal leaders, we get communities worth preserving. Our shamans provide the thread that connects the generations.

It is time for people in communities all across rural America to follow the thread backwards, and study again the lessons learned by our elders. We need those lessons, not just the liturgies and myths we often recite, because our rural communities are changing rapidly. Some have lost their reasons for existing. Some aren't coping with new technologies and global markets.

Our ancestors showed us how to adjust and cope with change. They built rural communities out of prairies and forest clearings. They fought for statehood to create better links to the outside world. They worked for land grant institutions that provide what in today's times are called infrastructure: transportation systems, education and research. They built cooperative, private and public finance institutions to transfer capital back to the frontier. They started mutual insurance companies to protect their assets and their crops. They started marketing and supply companies to move goods and commodities in and out of their communities. And they formed municipal and cooperative power utilities to string power lines across the land and bring new technologies and better standards of living to small towns and farms. Their legacy contains the tools to make changes, to pool resources, to revive rural America and rebuild communities that will flourish in the Twenty–First Century.

As noted in Chapter 1, the Northwest region of the United States has the cultural heritage that allows for new generations of cooperative businesses and for communities to work collaboratively to revive their agricultural and resource–based local economies. [2–1] There are models, discussed in later chapters, that would work in any community, in any region of the country. I hope rural community leaders explore the works of their elders, the native American leaders and immigrant entrepreneurs and laborers, who built the frontier. It's time to rebuild it, to revive it: the tools are available.

This book becomes a call for a cooperative revival in rural America. Not because private investment should be discouraged or that industry can't be lured into rural communities. The call for a cooperative revival comes from the assumption that government assistance to rural America is in decline, and that most rural communities will have neither the well capitalized local entrepreneurs nor the outside investors to create manufacturing jobs and raise the value of area commodities through local processing – the main sources of development capital.

There is a danger that our leaders, our shamans, may be only looking back, trying to preserve what was and cannot be again. To use the tools at rural America's disposal as we begin a new century, it is helpful to explore what has worked and is working, both in the United States and other parts of the world. [2–2] Most importantly, community leaders must look at what is coming at them from the top down, not from the ground up. For most rural communities, the task ahead is finding options and opportunities to fit into an integrated global food system that is changing rapidly and is being imposed on food producers.

ಐಐಐ

Marigold Foods is a fairly typical, mid–size American food company that is making adjustments to survive and flourish in the new world food system. To start with, it is Dutch owned, not local investor owned, so its American brands are among the nearly 25 percent of all major food brands on supermarket shelves that are foreign owned. And like nearly all food companies of any size, it was ready to expand and become a multinational food company in the 1990s.

The company began seeking strategic partners abroad whose existing presence in new markets could help Marigold avoid startup costs for both production and marketing. A 1993 joint venture partnership with National Dairies at Sydney, Australia, was started to allow Marigold to take its Kemps brand of Yo–J yogurt and juice drinks "down under" to an affluent market of nearly 20 million Australians.

With this arrangement, the Minneapolis food company joins the ranks of North American and European food firms that are spanning the globe and expanding at home with the help of joint venture partners. "What we're doing is following the new model," explains Marigold President Jim Green. "We don't want to build plants in Australia." [2–3] So Marigold, a modest sized player at the time in the world food market with annual sales of about $350 million, is following the same path for growth being taken by such food and agriculture industry giants as Europe's Nestle, Unilever and Grand Metropolitan, and America's Cargill, Archer Daniels Midland, PepsiCo and Kellogg.

The Marigold joint venture in Australia was scheduled to start producing Yo–J and distributing it to supermarkets by early 1994, said Mr. Green. And shortly thereafter, Marigold would start looking for another partner in the United Kingdom or somewhere in Europe. "We've become the biggest piece in a very small pie in the U.K.," he said. [2–4] The Kemps brand had taken about 50 percent of the U.K.'s frozen yogurt market, explained Bob Scheisel, vice president for marketing. Though starting small, with about $4 million a year in initial sales, the early success demanded Marigold aggressively develop its U.K. market.

The yogurt dessert products were being made at the company's frozen products plant in Rochester, Minnesota. That was a long, costly haul to London supermarkets for low–cost, novelty food products. For that market to grow as Marigold anticipated, the company would need a production partner much closer to its customers.

"You can bet on it," said Mr. Scheisel when asked if production would be taken to the British Isles. So Marigold went searching, looking to shift from one strategic alliance to another. Its employees at Rochester made the frozen yogurt desserts from milk supplied by southern Minnesota and northern Iowa farmers through their Associated Milk Producers Inc. cooperative. The dairy farmers will continue to supply milk to the Rochester plant, but Marigold will work to expand more serviceable domestic markets or use Rochester frozen yogurt to open yet another overseas beachhead.

Marigold was thus becoming part of a "realignment of the world's food system," explains Dwayne Andreas, chairman of Archer Daniels Midland at Decatur, Illinois. [2–5] He would know. His ADM business units, more than any other American based company's, have paved the way for joint venture partnering.

"We're seeing a new division of labor in the food industry," says Mr. Andreas. [2–6] "We've learned we don't have to do everything; we can have partners who do things better than we can, and we can do things for them better than they can do (for themselves)." Consumer food companies such as Europe's Grand Metropolitan and Unilever, he said, and American–based General Mills, Borden, Proctor & Gamble and Pillsbury are focusing more on their manufacturing and marketing strengths. "These people have the trade names and the shelf space (in supermarkets). They know how to deal at the consumer level. Advertising is vital."

In the middle of the realigned food chain are agricultural processors, Mr. Andreas said, with ADM and Cargill prime examples. "But there was a time when food companies thought they had to mill their own flour, process their own cooking oils and access all their grains and supplies. We (ADM and Cargill) can process their stuff better than they can – it's a matter of basic economics."

On the ground level, meanwhile, ADM has chosen to form joint ventures with farm cooperatives in North America and Europe to work cooperatively in trading, marketing, and, to some extent, feed milling and processing. "There are things the co–ops really do well, such as originating grain and raw materials," he said. "And they are on both the buying and selling sides for feed with their members. We have a natural relationship with them (because) we have 160 processing plants, 2,000 barges, 100 ships on the high seas at any given moment, and several hundred thousand trucks picking up grain and delivering to plants and export locations."

In total, Mr. Andreas explained, ADM has joint venture relations with seven U.S. and seven European farm cooperatives. Among venture businesses, ADM shares ownership of A.C. Toepfer International, a grain trading firm at Hamburg, Germany, that accounts for 50 percent of the agricultural trade in and out of Eastern Europe and

20 percent of the world traffic in feed ingredients. ADM owns 50 percent of Toepfer. Prominent American partners include Harvest States Cooperatives in the Northwest states, Growmark in Illinois, and Gold Kist at Atlanta, the Southern farmers' cooperative that also supplies peanuts to ADM oil processing plants.

These joint venture relationships helped ADM report sales of $9.8 billion in its 1993 fiscal year. The whole network of cooperating companies – co–ops, private ventures and public stock companies – had annual sales of $70 billion that year, Mr. Andreas said.

Such success with venture partners wouldn't go unnoticed at food and agribusiness companies at home and abroad. Even privately-owned Cargill Inc., the American trading and processing firm with $50 billion in annual sales, has moved into the partnering act. In the spring of 1992, for example, Cargill opened Saferco Products Inc., a $300 million nitrogen fertilizer joint venture in Canada at Belle Plaine, Saskatchewan. Its partner is the Crown Management Board of Saskatchewan, a body akin to a U.S. state investment board.

Though Cargill built itself by plowing back working capital and profits into plants and business expansions, it is likely to form more joint ventures in the future. "Partnering is the industry buzzword of the 1990s," says Cargill spokesman Garland West. [2–7] It is also a logical extension of traditional business relations, adds George Dahlman, a highly respected food industry analyst with the Piper Jaffray brokerage house. The food industry has always had partnership relationships between wholesalers and retailers, suppliers and manufacturers. Mr. Dahlman says the more formal business ties built through joint ventures are a spillover from general manufacturing. "It doesn't seem so unusual when you look at other industries." [2–8]

That's what officials at General Mills thought, too, when they formed Cereal Partners Worldwide with Nestle S.A. of Switzerland in 1991. "We were surprised when the world business press made such a big deal out of it," recalls Craig Shulstad, public relations manager at the American cereal, restaurant and consumer foods company. But it turned out no two leading consumer foods companies had previously created a new entity to manufacture and sell their major branded products in new markets. Cereal Partners Worldwide, based at Morges, Switzerland, near Nestle headquarters, combines General Mills' Big G cereal expertise with Nestle's Europe–wide marketing experience.

Nestle is the world's largest food company; it had 1993 gross revenues that translated into $33 billion. General Mills is the second largest breakfast cereal manufacturer, behind Kellogg, and had total food business revenues of about $12 billion in the same year. Within two years after starting production and marketing in Europe, Cereal

Partners expanded to Mexico and Latin America. It was on pace to reach $1 billion in sales by 1996.

This early success with a major joint venture partner was quickly duplicated. General Mills formed Snack Ventures Europe at Maarssen, the Netherlands, in 1992 with Dallas–based PepsiCo Foods. It, too, is an effort to weld the two food companies' strengths. Both companies had snack food products in European markets. General Mills was stronger in Northern Europe while PepsiCo had successfully penetrated Southern European markets in Spain, Portugal, Italy and Greece. Snack Ventures now sells both companies' products throughout the continent.

More typical of overseas business arrangements, General Mills also has a joint venture in Japan with Jusco, a large Tokyo retailing company. They jointly own Red Lobster Japan, which in turn operates 50 Red Lobster restaurants in the Tokyo area. Meanwhile, International Multifoods and Land O' Lakes are both venturing into Mexico and its fast growing economy. Multifoods owns a minority stake in two food processing companies, and Land O' Lakes started a joint venture with a Mexican food cooperative. Farmland Industries of Kansas City formed a wholly–owned Mexican subsidiary company in 1993, also with the objective of building inroads to the growing Mexican consumer market.

On the rebound from Europe, BioTechnica International at Overland Park, Kansas, has sold a controlling stake to Groupe Limagrain Holdings S.A. of Chappes, France. Dr. Charles Baker, president and chief executive of the American genetics and biotechnology company, described the deal as a "strategic alliance." "This alliance with Limagrain gives the combined companies access to additional technology and resources to grow and take advantage of future opportunities," he said. [2–9] Limagrain would roll its Shissler Seed Company at Elmwood, Illinois, into BioTechnica while it effectively took control of the latter. Groupe Limagrain, which earlier bought Shissler farm seeds and American vegetable seed companies Ferry–Morse and Advance Seed, was the world's third largest seed company at the time of the deal with 1993 sales of about $550 million.

Quietly, but in some cases more dramatically, Dutch dairy farmers have been expanding into North American food and agriculture markets through their Campina Melkunie cooperative. They now have operations in 130 countries and own, among foreign holdings, Deltown Specialties at Fraser, New York; DMV Ridgeview at Whitehall, Wisconsin, and DMV USA, at La Crosse, Wisconsin. They make specialty dairy products, dairy–based pharmaceutical products and industrial products from dairy components. [2–10]

None of the international business moves surprises Mr. Andreas. A native of Worthington, Minnesota, he formerly worked in a family soybean processing business and was an executive at both privately owned Cargill and a predecessor cooperative of Harvest States before taking command of publicly owned ADM. Throughout his agribusiness and food industry career, he has been one of America's leading advocates of soybean processing and the use of soy products. More recently, he has been one of the nation's leading advocates for developing ethanol and other new corn products. His soybean advocacy led to new business ties with Pillsbury Company, though it wasn't a formal joint venture. Pillsbury has introduced ADM's frozen meat substitute patties to groceries under the Green Giant label. These high protein replacements for hamburger, called Harvest Burgers, were test marketed in nine U.S. cities before Pillsbury began national distribution to about 50 percent of American supermarkets in 1994.

The interlocking relationships in the world food industry continue to expand. Pillsbury, now a North American property of London–based Grand Metropolitan Plc., has formed a joint venture business in Mexico called Pacific Star to distribute its refrigerated and frozen food products south of the border. The joint venture came after the Minneapolis food company bought 49 percent of Pacific Star de Occidente, a privately owned food distribution company in Mexico. Pillsbury will likely continue the business practice and become a major joint venture player in the United States and abroad. Grand Met, its parent, has emerged as one of the world's largest food and beverage companies with 1993 annual sales of $15 billion. Both the parent and the Pillsbury food unit operate joint venture businesses in Europe.

International linkages through joint ventures and acquisitions of specific assets from other food companies have become the mode of business expansion in the 1990s. In one spurt of acquisitions in the fall of 1993, for instance, Cadbury Schweppes Plc., based in the United Kingdom, bought A & W Brands Inc. for $334 million. While doing so, the Wall Street Journal and Forbes Magazine noted, Cadbury Schweppes also increased its stake in Dr Pepper/Seven–Up Companies to 26 percent, prompting food industry analysts to wonder if the British company was planning a takeover of the American soft drink firm, and whether it could do so under what remains of U.S. antitrust laws.

Unilever NV, the Dutch food company that rivals world leader Nestle S.A., expanded into the United States by buying the ice cream businesses of Kraft General Foods, a foods unit of Philip Morris Companies, for an undisclosed price estimated between $200 million and $300 million. Leaf, a not–so–sleeping giant of the North, has quietly bought so many different candy, food and consumer products lines around the world that it started in 1993 to announce itself to the consuming public. Not that kids or their retailers particularly care to know that Jolly Rancher and Milk Duds candies are products and brands of a Helsinki, Finland, food company; it may help the company in future growth, expansions and acquisitions if the world knows the Leaf name.

The American–based CPC International, meanwhile, bought controlling interest in Pfanni–Herke Otto Echart KG, a large potato processing company at Munich, Germany. In some instances, an American company has merely needed to travel across the state to work a deal for international expansion. Hershey Foods Corp. traveled the Pennsylvania Turnpike to buy the Italian candy business that had been established by H.J. Heinz.

Acquisitions will remain a popular route to business expansion when popular product lines and brands are involved. Manufacturing of consumer food products, however, is another matter. Even before Grand Met bought Pillsbury, the American firm expanded its Green Giant vegetable business to Europe by forming a joint venture. How Pillsbury did it is particularly relevant for American farmers and rural communities. [2–11]

Pillsbury teamed with farmers in the south of France and their Coop de Pau cooperative to grow and process Green Giant sweet corn. Ralph Hofstad, then the chief executive at Land O' Lakes, went to France at Pillsbury's request to help the farmers structure their processing business.

Mr. Hofstad is now retired from Land O' Lakes and working with groups in Russia to build joint ventures and a farm–to–consumer food system. [2–12] He said in a 1993 interview that Coop de Pau has expanded into other food processing businesses and formed additional joint ventures with European food companies. It is a model Mr. Hofstad uses in his work in Russia, he said. And it may prove to be an important model closer to home.

That's because Pillsbury began closing four Green Giant processing plants – three in Canada and one in California –in late 1993 as part of a $100 million reorganization of its North American vegetable business. A fifth plant, located in Ohio, was converted from packing vegetables to making Pillsbury's frozen pizza products. This means Pillsbury has started going outside its own Green Giant processing and canning factories to buy vegetable products made to company specifications, said Pillsbury spokesman Terry Thompson.

"These plants don't live up to the financial performance that companies such as Grand Met wants," says Dana Persson, president of Co–op Country Elevators at Renville, Minnesota. [2–13] A private

54

agribusiness company, such as a Cargill, or farmers, through a cooperative, can take "a longer view" if they are building their own markets and creating synergism with business ventures, he adds. Given these longer–term objectives, he says, "Farmers' expectations of returns on equity are less than the traditional institutional investors."

෯෯෯

That brings us down to the ground level. And it points up one of the strengths farmers and people in the countryside will have if they choose to carve out niches in the new global food economy.

Southern Minnesota farmers now grow and supply sweet corn, peas and other vegetables under production contracts with four nearby Green Giant packing plants. Wisconsin farmers also grow various vegetables they deliver to four Green Giant plants in their state. The Pillsbury unit of Grand Met still owns those plants. But farmers or a combination of farmers and townspeople should be prepared to take over ownership and management of those canneries in some future year, given current trends with the company. And not just the Green Giant plants, says Mr. Persson at Co–op Country. Del Monte, Bird's Eye and the other major vegetable canning and freezing companies will also be looking for ways to lower their capital expenditures in plants. "Their strength is in product development and marketing ... plants are just a drag on their balance sheets," says Mr. Persson. [2–14]

That being the case, farmers across America should expect the multinational food companies to change the way they contract for raw materials to make consumer food products. They won't have one way to access vegetables and process products in Europe and a different way in the United States. Other food producers, too, will find themselves adjusting to marketing changes brought by the internationalization of food companies. Orchard operators and marketers of Grannie Smith apples in South Africa and Chile already share markets and wholesale distributors with apple growers and marketing companies in Washington, Michigan and Ontario; and the Montana sheep and wool producers share lamb and textile markets with their agrarian colleagues in New Zealand and Australia. The global food system doesn't leave a lot of national or regional differences for the structure of agriculture.

Look at the linkages and market niches Upper Midwest dairy farmers share with their contemporaries in Holland: The Wisconsin dairy component products companies owned by the farmers of Campina Melkunie acquire fluid milk and processed dairy ingredient supplies from Associated Milk Producers Inc. farmers and their factories. Those same farmers supply milk to Marigold Foods' frozen yogurt plant in Rochester that was shipping consumer products to London. In Minneapolis, Marigold executives Green and Scheisel are considering a new joint venture with a dairy cooperative or private diary business somewhere in the United Kingdom to make yogurt from British or Irish farmers' milk.

Such a venture is almost certain to be sited in the British Isles, they said, and not on the European continent. That's because Marigold Foods has a new set of owners. Until 1993, it was a North American food company subsidiary of Koninklijke Wessanen of Amsterdam. That company has since merged with Bols to become Bols Wessanen – combining distilled spirits, wines, dairy and packaged foods under one company roof the same way the larger Grand Met has combined beverages and food. And with this merger, Marigold now has a corporate cousin in the European dairy foods business, Den Hartog, which is a Dutch ice cream company. Because of that company's European base, Mr. Green said Marigold will most likely look to Asia and the Americas for future expansion of its yogurt dessert business.

It should be noted that American food and beverage firms are buying more subsidiary units through mergers and acquisitions abroad than the foreign investment occurring in America's food and agribusiness industries. [Chapter Note A] In a real sense, enlightened business leaders throughout the world are building new corporate communities of shared interests to participate and grow in the expanding global marketplace.

Regardless where good companies go next with their finished products, or how they strike alliances and partnerships, processing plants will remain the lifeblood of rural communities. They provide jobs in rural areas when the technologies of modern farming no longer require large numbers of farm laborers to produce most crops. They provide nearby markets that can either enhance the value of area grown crops or save farmers the transportation costs they pay to ship commodities to distant processing plants. And equally important, though this gets scarce mention by leaders of major farm organizations, they provide part–time and full–time jobs for farmers and farm spouses when farm income isn't sufficient for family needs and when farm chores no longer demand full–time attention.

The latter reason why rural–based plants are important came to light in the mid–1980s when American farmers were still deep in the 1982–1987 farm depression. Some Farm Credit System banks were brought to the verge of collapse. Even large Chicago and California–based money center banks were weakened by the farm financial crisis, bringing close scrutiny of banking operations by federal financial regulators. [2–15] Rarely had agricultural finance been so closely monitored and analyzed as billions of dollars in farm debt began falling on urban–based institutions.

It was during that time when agricultural economists Emanuel Melichar of the Federal Reserve Board staff in Washington, D.C., Michael Boehlje, now at Purdue University in Indiana, and Neil Harl at Iowa State University began making the broader public aware of the enormity of American farm debt. They used data that was already available from Federal Reserve Bank reports and from financial reports assembled by the U.S. Department of Agriculture's Economic Research Service. What emerged from their published papers and public speeches was a picture of a radically changed American farm structure.

To simplify their reports, as they often did to help public understanding, they found a three–tiered structure of American farms in which the largest one–third was producing almost two–thirds of America's food and farm fiber. The middle–one third of farms produced nearly a third of American farm production, but it held a disproportionate high amount of farm debt that would be difficult to serve with anticipated farm revenues. The smallest one–third of farms were variously called "hobby farms" – or recreational farms and land holdings – or were rural residences at which some farming occurred while one or more people commuted to nearby jobs in towns, mills, mines or forests. These other positions produced most of the small farm families' income, and the farms' contribution to the nation's food supply was minimal. [2–16]

By the mid–1980s and the darkest days of the American farm depression, the economists were pointing out to anyone who would listen that the United States had between 600,000 and 900,000 farms, out of about 2.5 million total farms, that were serious contributors to the nation's food and fiber supply. For the vast majority, farming was indeed a way of life, not a profitable, productive business.

Statistics show the farm depression effectively ended in 1987, though not on all farms. The next few years brought record and near–record net farm income to the farming sector. Total U.S. farm indebtedness fell from about $200 billion in the early 1980s to $138 billion in 1990. [2–17] Financial institutions, and federal and state farm programs helped make from $50 billion to $75 billion in farm debt to disappear or get refinanced on more manageable terms and interest rates. But we now know this process was a painful ordeal that exacted a heavy price from the mid–sized group of farm operators. Some, who could get additional financing, expanded and joined the ranks of the large farms. More sold farm assets to better manage debt, thus slipping into the small farm category. Even more of these mid–sized farm operators, it would appear, left farming altogether. [See Chapter Note B]

ଌଌଌ

By 1992, the structure of America's farms bore little resemblance to the "American Gothic" image of the simple living, hard working farm couple, struggling against the elements and unpredictable markets to make a living for themselves while preparing a farmstead to turn over, in time, to children and their families. Janet Perry and Bob Hoppe, economists with USDA's Economic Research Service, found in a study of 1992 farm finances that 1.7 million of America's 2.1 million remaining farms were small farms, and that 90 percent of American farm households' average income of $40,068 in 1992 came from other, off–farm sources. [2–18] "Most off–farm income of farm households comes from wages and salaries or from off–farm businesses. In two–thirds of farm operator households, either or both the farm operator and spouse earned off–farm wage and salary income. Other sources of nonfarm income include interest, dividends and Social Security," they wrote. [2–19] As a result, Economists Perry and Hoppe have developed their own three–tiered structure of American agriculture.

Starting at the top, they define only 24 percent of all U.S. farms as "commercial" farms in which the operator reports farming as the major occupation and the farm generates at least $50,000 in gross annual revenues. They account for most farm production and receive 75 percent of direct government farm program payments, but the largest commercial farms still gain about 30 percent of farm household income from nonfarm sources. Smaller commercial farms receive about half their total household income from off–farm sources.

The economists defined the second, or middle group, as "viable noncommercial farms" that don't meet one or both of the commercial farms' criteria, but have household income of at least $15,000. These farms now account for 54 percent of all farm households. They lost an average of $817 on farming operations in 1992, but off–farm income provided average household income of more than $50,000 for these operators.

Finally, there is a bottom category of "low income farms" that account for 22 percent of all farms. They also account for a lot of America's rural poverty. The economists defined this category as farms lacking one or both criteria of commercial farms while producing household incomes of less than $15,000.

The demographics of this latter group is a particular threat to the economic health of rural communities that are distant from major urban centers. Low income farmsteads far from metropolitan areas are not likely to be turned over to a new generation of farm operators who will join the ranks of viable noncommercial farmers. If these farmsteads continue to be used as rural housing, they will most likely perpetuate rural problems with poverty. It can be safely assumed that a change of occupants, or abandonment, is coming:

"Operators in this group were older and had less formal education than other farm operators, limiting their off–farm opportunities. Fewer farm operators and members of these households reported performing off–farm work than commercial and viable noncommercial farm operators and household members. Most low income farms are located in the South (46 percent) and in the Midwest (35 percent).

"The situation of these households raises the question of why they remain if they are not making a living at farming. Some obviously stay because they prefer a rural lifestyle. Cash requirements to pay off–farm debt and cover living expenses are generally low for this group. Alternatives to farming, such as moving into a metropolitan area, may be less financially rewarding, given their limited education and training." [2–20]

ဆဆဆ

We've had so many shamans, so many wise men and women, to guide us over the years. Yet, too often, we fail to listen well enough to their message. We overlook what our ancestors did to build communities and build an infrastructure that let them reach distant markets. They reshaped their frontier, but didn't make it disappear. Now, the frontier is everywhere in the global market that remains beyond the reach of our commodities and products. But too often, we sit around at agricultural meetings splitting hairs over such subjects as whether farming is a business or a way of life, whether federal farm policies should support producers or production, and whether there is a difference between welfare and public support for private business that is – or was – deemed to be in the national interest.

Considering the demographic statistics mentioned above, it's probably fortunate for agriculture that nobody else pays much attention to these continuing philosophical debates. These questions have joined such parlor mind games as pondering the sound of one hand clapping, or questioning whether a falling tree makes a sound when no one is in the forest to hear it. A quick look at U.S. federal farm policy shows an obvious compromise has been reached over the past five decades: We support a shrinking number of large, well–capitalized farm enterprises while pretending to support small farmers for whom farming is a negligible source of income. Farming remains both a business and a lifestyle, but it's primarily a preferred lifestyle for nearly two–thirds of the American people living on farms.

America isn't yet ready to hold a meaningful public discussion about rural welfare. Are there better ways to support the rural poor than through a trickle–down farm price support mechanism that doesn't trickle down to most small livestock, vegetable or specialty crop producers? Such an honest discussion would require us to forget or ignore the myths about agriculture that have endured from the earliest days of our independence.

America probably isn't ready for a meaningful discussion on the future shape of farm income and production policies, either. Congress will write a new, multi–year federal farm program in 1995. But farm groups were preparing for that debate by looking for loopholes in the North American Free Trade Agreement and the Uruguay Round international trade agreements that were concluded earlier. Rather than working on drafts of sweeping new farm legislation that combined both economic and social objectives, most farm groups were seeking ways to continue their pet farm and commodity programs from past farm bills. Farm policies evolve, shaped by external pressures such as international trade, trade agreements and U.S. federal budget deficits. It is clear, however, as we approach a new century, that the future health and vitality of America's farms and rural communities won't depend on price supports for cereals, feed grains, oilseeds, fiber or dairy commodities.

The declining farm population, and the configuration of farm demographics, should tell everyone in agriculture that public support is about spent for big government payments to a few thousand large commodity producers. A discussion at a joint meeting of the North Dakota and Minnesota wheat growers organizations in the fall of 1991 revealed a general resignation to the changing times. How can farmers expect continuing support from a government that was allowing its big, populous cities to decay? Farm advocacy groups would argue that farmers merely need to elect a "better," or more responsive government. It's not that easy. Agriculture was practically ignored in the 1992 presidential election, and future elections won't hinge on the farm vote, either.

Congress and the administration began placing caps on spending for farm programs in the mid–1980s. Within a year of the 1992 election, smaller programs, such as those for wool and honey, were being phased out. Anyone who has watched farm policy debates from 1985 on has to wonder if there really is government support for major farm commodities. A strong argument could be made that commodity–based farm programs remained in law only to serve as a negotiating chip in the seemingly endless Uruguay Round of trade talks. Those talks have ended; our farm program's threat to world markets has ended, too.

ಐಐಐ

While public support for farm programs is declining in general, the federal farm program and related public policies are running into regional obstacles as well. In 1973, for instance, dairy cooperatives, dairy farmers and food companies banded together under the good offices of the National Milk Producers Federation to forge a consensus on dairy policy for that year's farm bill debate. [2-21] Twenty years later, the federation and large, multi-regional dairy cooperatives were awkwardly dealing with federal milk policy issues, giving their members a lot of slack to fight for their own agendas.

Farmers and dairy interest in Wisconsin, Minnesota, Iowa and North Dakota, for example, were waging battles in federal courts trying to strike down federal policies that boost farm milk prices the farther one travels from Eau Claire, Wisconsin. Those policies served an import public health purpose in the days before refrigerated transportation; they encouraged production of fresh milk for local markets in areas where farms were being built in deserts and swamps. Now, they continue to promote dairy production in deserts and swamps that are generally considered to be unnatural habitat for Holstein cows.

How bizarre is that? Former Minnesota Agriculture Commissioner Jim Nichols and other northern liberals who generally favor an activist government involvement in agriculture and markets, became the champions of free trade in milk and sought to end to federal distortions of the domestic milk market. Former U.S. Agriculture Secretary Clayton Yeutter, best remembered as the U.S. Trade Ambassador who sought an end to all agricultural production and trade subsidies under the General Agreement on Tariffs and Trade, became the unlikely defender of the domestic regional trade barriers and subsidies.

We have many such distortions and unnatural influences affecting what gets produced, and where. California replaced Wisconsin as America's leading dairy state in September, 1993. [2-21] Would dairying be more than a cottage industry in California if it weren't for Teddy Roosevelt and the Reclamation Act of 1902? Would California be a leading producer of any farm commodities, with the possible exception of citrus crops, if it weren't for the Reclamation Act? Farmers have managed to be priority users of California water that the act developed at great public expense.

Some people may ask why water for cows or rice is more important than water for people. People living outside of California might still conclude this water arrangement isn't all bad; better the farmers get the water than the real estate developers who would build even more housing tracts on the San Andreas Fault. But without the subsidized water, would former Agriculture Secretary Yeutter, have needed to spend so much time trying to get the Japanese to lower their trade barriers and accept imports from California's low–cost but desert–grown rice crop?

Ed Lotterman, an economist with the Ninth Federal Reserve Bank of Minneapolis, offers more evidence of how water, farm and public lands policies can distort agriculture within the domestic market. [2–22] His ancestors were vegetable growers in northern Illinois who provided fresh produce for the large Chicago market. That business ended with development of California water and refrigerated transportation. The Lottermans moved to southwestern Minnesota where they operated a diversified farm.

In 1962, he recalls, the family leased pastureland to an area sheep farmer using a sophisticated but common agricultural formula employed by the U.S. Departments of Agriculture and Interior on public lands in the West. In 1993, when Congress and the Clinton Administration were considering raising grazing fees on public lands, Mr. Lotterman discovered the Western public lands fees were identical to what the Lottermans charged their neighbor 31 years earlier. Meanwhile, the rate per animal unit for renting Midwest pasturelands had increased from four–fold to five–fold in the interim, reflecting costs of land ownership and the land's value for competitive farming purposes.

One can't say subsidizing Western sheep and cattle ranchers on public lands by charging inexpensive grazing fees is all bad, either. Clearly, there are environmental benefits for the broader American public from putting grazing animals on some of this land in the absence of wild herds of bison, elk and similar animals. But I don't buy the "divine right of agriculture" arguments offered by conservative Wyoming and Montana ranchers who refuse to recognize their cheap grazing fees are public subsidies. [2–23]

Take a summer vacation drive down Interstate–35 some year, from Duluth, Minnesota, to San Antonio, Texas. The rolling countryside of western Wisconsin and southeastern Minnesota, stretching out from the freeway, are natural grazing lands. So are the "bluff country" lands of northeast Iowa and southwest Iowa, the hill country above Kansas City in Missouri, the Flint Hills of Kansas, "Green Country" in Oklahoma, and the hill country stretching down through central Texas. Cows –

whether dairy or beef breeds –and sheep should be grazing those hills. Plows shouldn't break those grasses to grow cereal grains, corn or soybeans. That is happening, with about the same environmental results as could be predicted for plowing up the mountain pasturelands of Pennsylvania, Upstate New York and Vermont. Still, we have public policies in place that have chased the sheep out of the Midwest hills and now threaten to do the same to dairy cattle.

Other examples could be offered. Undoubtedly, there are combinations of farm, trade, environment and other public policies working at cross purposes in all regions of the United States. An agricultural economist at Tuskegee University in Alabama and I exchanged notes in the early 1980s about the impact of modern farming practices on black farm ownership in the South. Double cropping, the practice of raising a winter wheat crop that would be harvested in spring, followed by growing a second crop of soybeans on the same land, was changing the economies of scale for crop farmers in the Deep South states. Which farmers would gain access to the investment capital to expand? Which farmers would be displaced? Tuskegee was exploring these questions when my acquaintance moved to Southern University at Baton Rouge, Louisiana.

My interest in Southern agriculture was more cultural than economic at the time. A newspaper colleague, Walter Middlebrook , and I had several conversations about the farm crisis of the 1980s and its likely impact on black farmers. Neither Mr. Middlebrook, who moved on to USA Today, nor I did any deep research on the subject. But it was our unscientific view that black farm families in the South, like white and Native American farm families in the Midwest and West, were producing a large number of teachers and theologians as well as food and feed crops. Moreover, farm families tend to be two–parent households. Try to find an educator or social worker anywhere in America who doesn't see linkages between two–parent households and student achievement and youthful behavior. Such homes do provide sociological benefits for children and communities.

As time marched on, my interest in Southern rural life took on economic curiosity as well. The economies of scale have so changed in the Midwest and West that we now see rural populations that bear striking resemblance to rural dwellers we historically associate with the Appalachian states and Deep South. We have an entire new generation of rural poor living on farmsteads that simply provide inexpensive housing. If these people produce as much as $1,000 worth of local produce or feed a couple of steers for the local meat market, they are counted as farmers under government definitions. They are simply rural residents or are among the low income farm households identified by the USDA economists cited earlier in this chapter.

This is where local community investment and a new generation of cooperative businesses come in. Ways must be found to raise the value of local area farm commodities and natural resources that will, in turn, provide meaningful jobs and income opportunities. The new division of labor in the world food industry, as outlined earlier by Dwayne Andreas, offers a window of opportunity for communities willing to stake their communities' future.

The decades of the 1970s and 1980s saw small grain producers bought out or pushed aside by expanding farmers. The 1990s are seeing small livestock producers retiring or leaving the land, turning their farmsteads over to rural dwellers or hobby farmers who commute to jobs in nearby urban centers. Technologies and sciences have changed agriculture in ways that don't keep a majority of farm households gainfully employed on the land. And government policies haven't arrested the trends: If anything, the policies have sped the transition by rewarding the production of expanding farmers.

The day for benevolent governments to intervene in markets and artificially boost farm income has about run its course. Trade agreements and federal budget deficits won't allow it much longer. Federal farm policies don't reach the people with the greatest needs, anyway; nor was that the policies' intent. Regional, state and community actions must fill public policy and infrastructure voids left by a retreating federal government. Rural Americans must start over – just as their ancestors, the pioneers, did – to give their farms and towns reasons to survive and prosper in the next century.

Chapter 2 FOOTNOTES

1. The Northwest states used in this book range from Wisconsin on along the Canadian border to Washington and Oregon. Its usage is historical; the region shared a common development involving immigration patterns, transportation and markets.

2. Egerstrom, Lee. "Rediscovering Cooperation." A paper presented to the Regional Directors Workshop, Jan. 7, 1993, and published by the Minnesota Association of Cooperatives.

3. "Growing by Joint Venture." St. Paul Pioneer Press. Oct. 18, 1993.

4. Ibid.

5. Andreas, Dwayne. From telephone interview conducted in September, 1993.

6. Ibid.

7. "Growing by Joint Venture."

8. Dahlman, George. Interview conducted for this chapter.

9. BioTechnica company announcement.

10. Egerstrom. "Rediscovering Cooperation."

11. From interviews. Cooperative processing joint ventures are discussed in more detail in later chapters.

12. See Chapter 6.

13. "Growing by Joint Venture."

14. Ibid.

15. For a comprehensive account of the Farm Credit System banks' financial difficulties during the 1980s, see: Sunbury, Ben. The Fall of the Farm Credit Empire. Iowa State University Press. Ames, Iowa. 1990.

16. From public speeches and papers presented at farm finance conferences during the 1980s agricultural depression. One particularly helpful paper was: Melichar, Emanuel. "Agricultural Finance: Turning the Corner on Problem Debt." Published by Board of Governors, Federal Reserve System. Aug. 26, 1987.

17. Ibid.

18. Perry, Janet, and Hoppe, Bob. "Off–Farm Income Plays Pivotal Role." Agricultural OUTLOOK. Economic Research Service, U.S. Department of Agriculture. November, 1993.

19. Ibid.

20. Ibid.

21. I reported from Congressional hearings and floor debates on the omnibus farm bills of the 1970s and monitored Congressional and administration actions on successive bills during the 1980s and 1991. It is my observation that commodity groups and regional farm interests are played against each other. The 1973 and 1975 farm bills were memorable in that dairy interests were unified when dealing with Congress and withstood opposition to their programs from the Nixon and Ford administrations.

22. October monthly Dairy Production Report, National Agricultural Statistics Service, U.S. Department of Agriculture.

23. Lotterman, Ed. Economist, Federal Reserve Bank of Minneapolis. From interview conducted in November, 1993.

24. My assessment, based on Western reactions to federal grazing fee debates in Washington as reported in Farm Bureau News, 1993 editions; and from report to members in the newsletter of the National Association of Wheat Growers, 1993.

Chapter 2 CHAPTER NOTE 2–A

Most of the international investing in food and agriculture companies cited in this chapter has gone unnoticed by most Americans. Employees who work for one of the involved companies, and residents of communities where ownership of local companies or plants has shifted abroad are exceptions since most such changes are met with some apprehension. But even employees of large corporations pay little attention when their company buys or partners with companies abroad if it doesn't affect their own work responsibilities.

Cowley Research & Publications, based in Amsterdam, the Netherlands, monitors cross–border merger and acquisitions each year for different industry sectors. In a report on the food and drink industry released in 1994, the Cowley organization recorded a total of 1,701 international merger and acquisition deals during the five years covering 1989 through 1993. Those transactions had a total value of $53.2 billion. American food and beverage companies accounted for 16 percent of the deals, or 273 cross–border investments, that totaled $15.1 billion, or 28.4 percent of the value of the transactions.

Among major deals cited by the Cowley report were Philip Morris' acquisition of the Jacobs Suchard chocolate and candy groups in Switzerland for $3.8 billion and Scandinavian–based Freia Marabou for $1.4 billion; PepsiCo's purchase of Smith's and Walker's potato chip business in the United Kingdom for $1.35 billion; Anheuser–Busch's $477 million purchase of 17.7 percent of Grupo Modelo, the largest Mexican brewery; Campbell Soups $223 million purchase of Australia's Arnotts snack food company; and Coca–Cola's $195 million buy of 30 percent in Femsa Refescos, Mexico's largest soft drinks company.

Cowley Research & Publications, which is a spinoff from the KPMG international accounting firm's database systems, is based at P.O. Box 3932, 1001 AS Amsterdam, the Netherlands.

Chapter 2 CHAPTER NOTE 2–B

Highlights of the Perry, Hoppe study, as reported in Agricultural OUTLOOK, November, 1993:

"Commercial" farms in 1992:
* 72 percent were in favorable financial condition; 22 percent of large farms and 26 percent of smaller–sized commercial farms lost money in 1992. Average household income of the large farms was $63,000 and average household income of the small commercial farm was $17,373. 61 percent of commercial farm households received majority of their income from farming, while small commercial farm households received about 50 percent of family income from off–farm sources. Average net farm worth of large commercial farms exceeded $1 million, and they received about two–thirds of federal farm price support payments in 1992.
"Viable noncommercial" farms in 1992:
* 54 percent of all U.S. farm households are in this group. The distinguishing feature of "viable noncommercial" farms is that households had off–farm wages, salaries and professional fees to – as some Midwestern economists call it – "subsidize their farming habit." For most households, these farms are preferred ways of life; the group lost an average of $817 from farming operations while households had average incomes in excess of $50,000. Net worth of these farms was in excess of $200,000, even though they contribute little to the nation's food supply and are more a hobby than a business.
"Low income" farms in 1992:
* These are the rural poor people who the American public rushes to support even though they do it by dropping large commodity program payments on the "commercial" farms. These people represent 22 percent of the farm households, and only one–third of them have favorable financial

positions with positive net incomes and low debts. Nearly half of these households are found in the South and a third are located in the Midwest. They lost an average of $7,334 on farm operations in 1992, reducing total on–farm and off–farm income to an average of $4,216 per household. These farm operators tend to be older and less educated than the general public, meaning these farms are not likely to be recycled to new operators; but the farmsteads may become housing for subsequent generations of rural poor.

A breakfast session at Knight–Ridder Newspapers. From left, the late Walter T. Ridder, Vice President Walter Mondale, former Knight Ridder Washington Bureau Chief Bob Boyd, and the author, Lee Egerstrom.

Photo courtesy of Stan Jennings

CHAPTER 3: OBSERVING CHANGE

"I tell beginning journalists they should never write about drought until they
move their families to high ground."
Don Muhm, former farm editor, the Des Moines Register.

"I learned from working in Florida, Mississippi, Maryland, and now
Minnesota, that there are some things all farmers have in common. They are good
at solving problems, but they don't solve equations."
Richard Levins, Extension farm economist.

Journalists are the paid observers who write the first line in the
equations social scientists use to forecast future events or explain
current happenings. The economists, political scientists and
sociologists write a second line, a comparison or contrast, then apply
mathematical logic. What takes shape is an equation, or syllogism,
from which a forecast or explanation evolves as a logical conclusion.
Eventually, historians grade all our papers to determine when we were
right and when we were just blowing smoke.

Professor Richard Levins, a witty agricultural and applied
economist at the University of Minnesota, has playfully reminded
colleagues that deductive reasoning is essential for forecasting and
analyzing, but English reads better than equations in the countryside.
[3-1] And Don Muhm, the veteran Omaha and Des Moines agricultural
journalist, reminds us that change is constant, news events are
transitory; tomorrow's newspaper may rewrite the first line for any
social formula. We all get blindsided by unexpected events.

Change, while constant, can either creep up or rush at people. This
has been observed by two veteran economists at the University of
Minnesota, Regents Professor Vernon Ruttan and Philip Raup, a
professor emeritus, who tell dramatic stories about social change
brought by use of the modern farm tractor.

Professor Ruttan is the world's leading authority on induced
technological change. He was introduced to the concept as a young
man when he was pulled off his family's Michigan farm for World
War II. The war created labor shortages on farms all across America. In
some parts of the Midwest, the labor shortage was made worse by an

outbreak of encephalitis that killed work horses. Horses were dropping even as family farmers were losing their helpers who could work the horses. "My father went shopping for a tractor when I got my 'orders'," Professor Ruttan says. So it went on farms across rural America, with mechanical tools eliminating the need for farm laborers by the time the war ended.

Professor Raup has witnessed dramatic social change abroad that was brought by the introduction of technology. He was asked by the United Nations' Food and Agriculture Organization in Rome to form a team of Western farm experts to help Ethiopia develop its agriculture. The project started in the late 1950s with a team of American, Canadian and European economists, agronomists and agricultural engineers. A palace representative of Emperor Haile Selassie met them at the airport at Addis Ababa, introduced them to a prominent land baron who was a friend of the emperor, and they were whisked off into the countryside on their initial inspection tour. "We would stop by a house in every town we visited. The baron would introduce us to his wife and children. Then we'd go see the fields and storage facilities," Dr. Raup said.

Nearly two decades passed and the United Nations called again. The economist assembled another team, including a Canadian who was part of the earlier mission. The same palace emissary greeted them, and the same land baron. But this time the tour of the countryside was all business. "Finally, curiosity overcame us. The Canadian asked, 'Where are all your wives and children?'

"We weren't ready for the answer," Dr. Raup recalls. "The man said, 'I don't need them. I bought tractors.'"

 හයහයහය

Bystanders to history often observe anecdotes that make strong arguments about social and economic changes. For most people, these personal observations can be more convincing than the best developed equations and academic arguments. Often, however, the latter need the first–person accounts of the former. Who can't grasp the need for a tractor on the Ruttan family farm in Michigan when Uncle Sam needed the plowman's muscle? What horrible lessons Dr. Raup teaches us about the economic value of women, children and labor in some countries, about the impersonal nature of technology, and, without question, the reasons why Ethiopia was ripe for a military coup shortly after his last visit.

This chapter is like a reporter's notebook filled with anecdotes. It reaches into the memory of the author and others who witnessed moments of change that are tied directly or indirectly to points of argument found in other chapters. It is not an essay. It is a series of first lines assembled for readers to build a logical case for changing their rural businesses and communities. More pointedly, these events have convinced me that the world food system has changed, along with other markets and systems for life's basic raw materials. There is no turning back. Change must be accommodated.

෨෨෨

I stayed out of college and worked for my hometown weekly newspaper, the Kerkhoven <u>Banner</u> in Minnesota, during the fall quarter of 1963. That means I was living at home and working in a small town in west–central Minnesota when President Kennedy was assassinated. Like all Americans old enough to remember that event, I recall vividly where I was and what I was doing when the news spread.

Now, three decades later, a related incident of December, 1963, is also an important part of that memory. A corn farmer from near Benson, Minnesota, brought a press release to the <u>Banner</u> office announcing the discussion topic for that month's meeting of the Six Mile Grove Township Farmers Union. The farmers were to explore the future of President Kennedy's Alliance for Progress program that was designed to integrate Latin America's economies with the United States'. The U.S. government's motives for the program were easily understood; it was a defensive move to undercut Fidel Castro influence in the Western Hemisphere. Farmers were equally quick to see potential economic benefits from the program. Closer trade and political ties, and economic progress in Latin America, were certain to lead to more U.S. farm exports.

For the record, these farmers were justifiably concerned about the future of President Kennedy's Latin America initiative. Nothing much did come of it. The Vietnam War and crisis situations in other parts of the world stole attention away from the Americas. Thirty years later, the United States was still working on integrating hemispheric economies through a North American Free Trade Agreement, linkages with Caribbean nations and potential trade agreements with Peru, Chile and other South American countries. Some members of the Farmers Union aren't as supportive of freer trade as they were 30 years ago, having formed alliances with labor organizations and environmental groups that sought other agendas. But the biggest difference in 30 years for the federal government was in motivation: This time around the national interest was economic security, not defense, as the United States works to open borders for economic growth.

71

That incident also taught a 20–year–old man a rather basic fact of economic life: The most remote, small farms in rural America are directly tied to political and economic developments everywhere in the world, every day of the year.

జుజుజు

In early 1972, Al Eisele took me in tow and introduced me to Midwest senators and members of the House of Representatives. I had joined the Washington bureau for what was then Ridder Publications, which would later merge into Knight–Ridder Newspapers. Among our Midwest newspapers were the Aberdeen American–News in South Dakota, the Duluth Herald and News–Tribune in Minnesota, the Gary Post Tribune in Indiana, the Grand Forks Herald in North Dakota, the Niles Daily Star in Michigan and the St. Paul Pioneer Press and Dispatch newspapers in Minnesota. The bureau also wrote for five Ridder newspapers in California. To introduce me, Mr. Eisele explained that I would cover what the Midwestern lawmakers were doing on the job in Washington. More senior bureau members, primarily Bureau Chief Bill Broom and Mr. Eisele, would cover the political aspirations and presidential campaigns of the region's aggressive public figures.

What this meant was that an inexperienced cub reporter would write what government was doing to and for people while experienced people would write about political ambitions and posturing. This arrangement must strike social scientists and historians as crazy, but it is a pretty typical structure inside journalism. This arrangement also offered me the greatest learning experience of my life: When I was learning that military might wasn't buying the Soviet Union food security, senior reporters were analyzing Indiana Senator Vance Hartke's campaign in New Hampshire; when I was learning that food and energy trade are linked, and are the foundation of both the global economy and international relations, the experienced correspondents were wading into the day's big stories, such as California Congressman Pete McCloskey's challenge to President Nixon's renomination. Sen. George McGovern of South Dakota listened to Mr. Eisele introduce me, then offered some advice. I could do readers in the Dakotas, Minnesota and Wisconsin a service by writing about agriculture, trade, national resources and economic policies, he said. But doing so would have social costs, he warned. "You don't get invited to Georgetown parties to explain what's in the farm bill."

జుజుజు

72

As agriculture was beginning to emerge from its financial crisis of the 1980s, California's Silicon Valley technology industries slumped into a serious depression that would nearly destroy the U.S. microchip industry, wipe out large numbers of small technology manufacturers and force the consolidation of other firms into larger companies. Don Clark, a talented technology business writer with the San Francisco Chronicle, called one day to commiserate about the state of agriculture and tech industries. "I think technology has caught up to where agriculture has been," he said. "We have Japan trying to sell us tech that we don't need. We are trying to sell the same products to Europe, but it has its own. Nobody in the developing countries has any money left to buy anything."

Mr. Clark was correct. American agricultural trade hit the skids when the Federal Reserve System and central bankers in the industrialized nations decided to put a stop to inflation that had been nearly out of control since the early 1970s. A decade of fine tuning the economy hadn't worked, so the bankers got a bigger hammer. They tightened the money supply, which in turn raised the value of the U.S. dollar in international currency markets. What had been cheap American farm goods became extremely expensive food, after currency translations, and Third World customers either went shopping elsewhere or tightened their belts. The same thing happened to America's industrial exports, resulting in the 1981–1982 national recession. [3-2]

Silicon Valley took a pounding, as did agriculture. A recovery would come to the survivors in both, but not as quickly as it did to the general U.S. economy. [3-3]

ဢဢဢ

Shortly after the Soviet Union invaded Afghanistan, Agriculture Department spokesman Jim Webster called to tell me to stay by the telephone at my office in Washington. There would be press briefings later. "We're going to dump some tea in Boston harbor," he said. And before the day was over, they had. President Jimmy Carter imposed what become the most politically charged and economically inconsequential trade embargo since rag–tag colonists dumped the King's tea in Boston's bay. It is still cursed at farm meetings all across the land.

The embargo was limited to begin with; it was a ban on any grain sales in excess of a substantial amount of trade protected under an existing U.S.–Soviet grain agreement. The agreement spelled out a

73

minimum amount of grain and oilseed purchases, so U.S. farmers could make planting decisions with some certainty of a market. It stipulated that Soviet buyers would need U.S. Department of Agriculture approval to buy more than amounts allowed in the agreement. In theory, the agreement would give both Cold War trading partners a measure of security. The Soviets knew they could count on the United States as a reliable supplier for a large amount of farm commodities, and the United States knew the Soviets couldn't interfere with the agricultural and broader U.S. economy by jumping in and out of our markets unannounced. The latter was deemed important. Agriculture and its friends in Congress didn't want – of all things – an embargo like Presidents Nixon and Ford had imposed on Japan to protect domestic U.S. soybean supplies and the national economy from inflation. [3–4]

Any possible impact the embargo could have on U.S. farm prices was further neutralized when Mexico entered the grain market that year, buying more grain than anyone had forecast for excess Soviet purchases. Indeed, the United States was in midstream of setting back–to–back records for agricultural exports in fiscal years 1980 and 1981 – the two years affected by the embargo. [3–5] But grain prices began falling, and grain merchants and politicians were quick to blame the embargo for the fall.

Political candidates, with former California Gov. Ronald Reagan seizing the lead, did everything but promise the Soviets we'd send "K" and "C" rations to their troops in Afghanistan after the 1980 election. So unpopular was the Carter embargo that campaigns against it reached down to candidates for county board positions. The politics I could understand. Demagoguery is a fond political tradition and I had seen Republicans – far from Washington – victimized in the aftermath of the Watergate scandal. But I was among Washington reporters covering the Agriculture Department who were baffled by falling grain prices. Was the grain trade playing politics? Was it the result of market panics, like futures and options traders suffer when they see rain splashing on the sidewalks in downtown Chicago? Were other market signals driving down prices beyond the highly visible embargo?

This reporter's gut instincts would say, "Yes. All of the above." But clearly, there has to be an economic linkage between falling agricultural prices in the early 1980s with monetary policy when, as noted above and in Chapter 1, there was a linkage between monetary policy and trade flows for agriculture and manufactured goods. If so, then markets were adjusting to monetary influences – whether traders in the nation's commodity exchange pits realized it or not.

The central bank – the Federal Reserve System – had tightened the money supply to squeeze inflation out of the domestic economy. This, in turn, raised the value of the U.S. dollar in international currency markets. Starting in 1980, the value of farm commodities and other American exportable products began falling at home even as the their costs remained high to overseas buyers who needed to trade their currencies into scarce, more costly dollars to make the deal. Furthermore, the same inflation–fighting monetary policies that drove up the value of the dollar also drove up U.S. interest rates as customers for credit competed for access to capital. Those rates contributed to a bearish mood around commodity exchanges. Though Mexico was making huge grain purchases, traders knew Third World countries would hesitate about borrowing money from international banks to buy costly American goods. Moreover, Third World countries were already in hock to international banks and had to pay existing debts with more costly, scarce U.S. dollars.

The inflation fighters clobbered inflation, making America's raw material industries the shock absorber for the broader U.S. economy by stifling grain sales long after the embargo was lifted. They also closed Louisiana, Texas, North Dakota, Montana and Alaska oil wells, shut down Northern Tier mines and stopped logging in the forests.

Did Americans learn anything from the Carter grain embargo? Not if they blame it for economic events connected to monetary policy, not foreign policy. [Chapter Note A]

ಐಐಐ

In the fall of 1984, reporters from the Minneapolis Tribune (now the Star Tribune), the Rochester, Minnesota, Post Bulletin, the Country Today farm publication in Wisconsin and St. Paul Pioneer Press hosted a conference on trade issues for members of the National Association of Agricultural Journalists. It was a strange time, both at home and internationally. The polls showed President Reagan had re–election sewed up by somehow arguing that we could have government without paying for it and blaming the Carter grain embargo for depression in the nation's farm, timber, plywood, iron ore, copper and coal sectors.

Former Vice President Mondale was winning some faint praise for integrity from journalists and pundits, but American voters weren't ready to accept a liberal's call for fiscal responsibility. Americans were also ignoring that Socialists in Europe, led by the French, were berating the United States about getting its financial house in order while Conservatives, led by the UK's Prime Minister Thatcher, didn't see a problem with the United States subsidizing the global economy with deficit spending. We planners for the journalists' conference went to great lengths to avoid 1984 politics. We turned to G. Edward Schuh to be our keynote speaker. "Stop worrying about who's going to be Secretary of Agriculture," he told the writers. "People in agriculture should start worrying about who's going to be Secretary of the Treasury." [3–6]

Professor Schuh later became the director for Agriculture and Rural Development at The World Bank in Washington and has since become dean of the Hubert H. Humphrey Institute of Public Affairs at the University of Minnesota. He should not be blamed for anything this journalist or other farm and trade writers have written, but he should be credited for helping journalists include some attention to economic detail in their reporting on trade and political events.

ଛଠଠଠ

There was no night in eastern Saudi Arabia. Flames from the oil fields lit the ground from the border of Kuwait to Dhahran, the Houston–or Tulsa–like oil capital for the Saudi oil industry. I was about to learn how the economics of raw materials are the same, the world over

The flames below were a terrible waste of precious resources, insisted Robert Buckler, a researcher for the U.S. Senate Interior Committee, as we flew across Saudi Arabia. Mr. Buckler and I were traveling with Sen. James Abourezk of South Dakota during the Arab oil embargo against the United States following the October, 1973 Arab–Israeli War. The flames were from flares at oil wells, Mr. Buckler explained. Natural gas escapes when crude oil is pumped from the wells. The gas was a danger to oilfield workers and the wells themselves unless burned, or flared, off.

Most of that gas could be captured and piped away for use as clean burning fuel by municipalities, utilities and manufacturing plants. Such enormous quantities of an important, non–renewable resource shouldn't be merely burned as a safety precaution.

Clyde LaMotte, an energy writer now retired and living in Charlotte, North Carolina, remembers seeing East Texas oil fields aglow at night when he was a young reporter and editor at Austin and Houston newspapers. The flaring ended, he says, with the massive oil and gas pipeline building that was done for defense purposes during World War II. This development would have happened anyway, Mr. LaMotte insists. The domestic supply of natural gas was too valuable to be wasted, he said, although most Americans would remain

76

indifferent to declining domestic energy reserves until more recent decades.

The Middle East oil producing countries also needed to find uses for their valuable natural gas resources, Mr. Buckler observed as we traveled westward across the Arabian Peninsula. It would not happen immediately, however. Saudi Arabia and the Gulf States were already engaged in the most rapid transformation from a rural, agrarian and largely nomadic culture to an urban, developed and industrial culture in world history. Change was coming with incredible speed; perhaps faster than the desert cultural institutions and governments could accommodate, he warned. [3-7]

ဆဆဆ

Adjusting to Middle East economic and industrial development has caused stress and pain along the northern Schelde River waterfront at Antwerp, Belgium. Though Antwerp is best known worldwide as the largest diamond trading center, it is also home to one of Europe's largest oil refining and petrochemical manufacturing complexes. Refined and manufactured petroleum products are loaded aboard tanker vessels, which are a new generation of self-contained barge ships, and then distributed to European industrial cities that are served by the Rhine River navigation system.

By June, 1985, business had slowed significantly at the Schelde port from the refining and shipping volume of just two years earlier. Europe's economy was strong and wasn't responsible for the low port business. Increased imports of Middle East refined products were.

"The OPEC (Organization of Petroleum Exporting Countries) exporters don't want to ship only crude oil anymore," explained a Rhine River shipping executive. "We still get crude oil from OPEC. But not as much. They insist we take more of our petroleum as refined products." [3-8] This was reducing some waterways shipping volume by bypassing petrochemical and refining complexes such as Antwerp's. It was also eliminating some jobs at the petro ports and refineries.

I joined a crew on a Nedloyd Rijn en Binnenvaart Co. tanker that was loading naphtha for delivery to industrial plants at Cologne and Dusseldorf, Germany. Around midnight, after six hours of loading, we would begin a three-day cruise up the Rhine to the German factories. The time in port allowed a Nedloyd executive and myself a night on the town. A stop for coffee and brandy at a neighborhood bistro on the waterfront found us engaged in friendly conversation with waterfront residents and off-duty refinery workers.

Most of the refined petroleum products entering Europe were coming from North African countries, the workers said. It was an inevitable development, they conceded. Every country is looking for ways to increase value–added processing and manufacturing, they said. Building large petrochemical and refining complexes allowed OPEC countries to improve their export earnings, create industrial jobs for their people and use their natural gas supplies for their own energy needs.

ಬಬಬ

Greg LaMotte stood on the rooftop of his hotel in Riyadh, Saudi Arabia, watching the nightly cat–and–mouse games between Iraqi Scud missiles and American Patriot missiles above central and eastern areas of the Saudi kingdom. He was among Cable News Network (CNN) reporters assigned to the Middle East during their 1991 Gulf War with Iraq.

"It was dark. There weren't any oil wells flaring," he recalls, though he's heard his father, the former energy writer, describe the similar scenes in pre–World War II Texas. "We didn't see flames until we got into Kuwait where the Iraqis had started fires in all the wells before retreating."

ಬಬಬ

Employment and business activity has recovered at Antwerp's petrochemical complex in the 1990s, says Thierry Bechet, a diplomat with the European Community and former Belgian Foreign Ministry official. He had just toured the Schelde River waterfront before visiting Minnesota in the summer of 1993. [3–9]

Trade flows hadn't reversed, and Europe wasn't receiving a larger percentage of crude oil for its petroleum needs, he explained. Rather, the Antwerp port had adjusted and was doing more manufacturing of further processed petroleum products at the complex. The distinction between refining and petrochemical manufacturing had become clear; the OPEC countries are raising the value of their precious oil resources before it leaves their shore while industrialized customers were further processing into factory–ready and consumer products.

Parallels are everywhere in the North American resources industries. Miners and workers at mine sites now process low grade ore into taconite pellets ready for shipment to Pittsburgh steel mills. Hormel Foods closed its hog slaughtering operations at its new state–of–the–art meat packing plant in Austin, Minnesota, then leased those facilities to Quality Pork Processors, a Texas company, to kill and cut meat for Hormel workers to further process and pack into branded, consumer food products. In early 1994, Seaboard Corp. of Merriam, Kansas, decided to close its slaughtering operations at its Albert Lea, Minnesota, plant, while workers there were to continue making hams, bacon and sausage products.

Paper mills in the forests along the Canadian border are turning to other, independent companies for pulp supplies. This is especially so in the 1990s as paper companies are responding to public interest in recycled paper products; not every paper mill wants to invest in building a recycling plant at their mill sites. The trends are unmistakable, offering companies more efficient use of resources. More often than not, they can provide environmental benefits as well.

The most alarming parallel exists between the Mid–America farmers and the Middle East oil producers. The OPEC countries knew they would remain essentially Third World countries with under–employed populations unless they started doing more value–added processing of their natural resources. But all across the land, large numbers of America farmers choose to operate like affluent peasants, simply growing and raising crops and animals for others to use and add value. OPEC's oil industry has reacted boldly; American agriculture is still waffling over its future structure.

ॐॐॐ

Change isn't easily accommodated. It is often resisted. When it is, the bottom line usually means more pain and economic hardship than was necessary. Consider these examples:

Teletype operators in Rome fought a losing battle to preserve jobs for their skilled craft during the United Nations World Food Conference in 1974. They would refuse to send journalists' news stories unless there was someone at the receiving end who would answer the telephone and then connect the Rome overseas line to the American teletype machine.

Night after night, American journalists would explain that the telephone number written on the news copy was for a direct line to a machine. No one would answer. The machine would answer and start receiving. The Rome teletype machine could begin transmitting the story immediately following a tone signal. But the only way to get the operators to accept the tone signal and send the story was to stand there, from one to two extra hours after a long day, and watch the operator do it. "Nobody answered the telephone," the operators would explain each day.

Could any journalist doubt that someone would invent the facsimile machine?

ഔഔഔ

American automakers and autoworkers fought losing battles for most of two decades, recalls John Peterson, a veteran Washington correspondent for the Detroit News who reported on auto regulatory developments and trade. Now a public relations executive in St. Louis, Mr. Peterson says government policy was often shortsighted, but it was usually a reaction if not retaliatory response to auto industry obdurance.

"I don't think there was a any single benchmark, but there were several incidents that strung along to weaken the industry," he said. They included company attempts to smear consumer advocate Ralph Nader instead of addressing public concerns about auto safety; labor strikes during recessions, and constantly turning to trade protectionism instead of accommodating change. Longer term, Mr. Peterson said the industry's resistance to improving fuel efficiency and making smaller cars left large segments of the U.S. auto market open to Japanese imports after gasoline shortages and higher petroleum prices in the 1970s. "Deadlines for fuel efficiency and safety were imposed on the industry. They may not have been well conceived," he says. But it was an inevitable response to an industry that had stopped working cooperatively with lawmakers and policy makers on widely perceived public needs.

ഔഔഔ

Feb. 2, 1985   St. Paul Pioneer Press (headline)

FARMERS SHOUT DEFIANCE AT AUCTION

By Lee Egerstrom
Staff Writer

ORTONVILLE (Minnesota) – Sheriff Bud Haukos waited as long as possible.

Then, as a sheriff must, he sold the farm that had belonged to -----.
[3-10]

He read the legal notice of foreclosure and asked if there were any oral bids. It is doubtful anyone heard him.

80

More than 100 protesting farmers and sympathizers stood in sub-zero temperatures on the Big Stone County Courthouse steps Friday yelling, "no sale, no sale" in a chorus that lasted 30 minutes.

The sheriff's sale late Friday completed a wave of farm foreclosure auctions and farm protests that had swept across western Minnesota.

It had started in Glenwood when several hundred protesters gathered to shout down a scheduled 10 a.m. sheriff's sale at the Pope County Courthouse.

That sale never came off. The Travelers Insurance Co., holder of a mortgage on ----------'s farm, decided Thursday night to cancel the auction.

----------, a former Minnesota president of the American Agriculture Movement, said the insurance company has given him until March 18 to remain on the farm. That will buy time for the Legislature or Congress to act on farm foreclosure moratorium legislation, he said.

Later in the day, foreclosure auctions were conducted without incident by sheriffs in Lac Qui Parle and Chippewa counties.

In the Chippewa County sale at Montevideo, the Federal Land Bank bought the ---------- farm. Sheriff's deputies said there were no demonstrators.

------------ has a farm near Graceville in Big Stone County. The mortgage is held by the Federal Land Bank Association of Willmar.

The demonstrators occupied the Land Bank's branch office in Appleton and implored office manager ----------- to grant ---------- another 30 days.

----------, a University of Minnesota-Morris history teacher and leader of the Groundswell farm protest movement, said the 30 days would give legislators and the governor time to impose a moratorium on foreclosures.

(The office manager) was a captive in his office for most of the afternoon.

He insisted he couldn't order a delay in the foreclosure. Calls to ---------, his manager in Willmar, and to officials at Farm Credit Services in St. Paul failed to produce a delay.

At one point, (the office manager) tried to leave his office but was pushed back by the crowd. "Grab a rope," yelled one farmer. "Hand out his pants first," yelled another.

(The farmer), meanwhile, went to Fargo, N.D., in a last-ditch effort to file a Chapter 11 bankruptcy petition to reorganize his debts. The North Dakota Federal District Court declined to accept the Minnesota filing.

That put the future of the -------- farm on the Big Stone County Courthouse steps.

Sheriff Haukos held up the foreclosure auction for more than half an hour as protesters sought word from the Fargo court. They then tried in vain to get Gov. Rudy Perpich to declare a foreclosure moratorium by executive order, just as Gov. Floyd B. Olson did in 1932. The caller didn't get past Terry Montgomery, Perpich's staff assistant, who said it couldn't be done.

The protesters threatened to keep the sheriff and his deputies in their offices so Haukos couldn't complete the auction.

"Don't do that to me," Haukos said. "Remember I'm not your enemy. I've got a job to do - to uphold the law."

There were many tense moments. Several protesters shouted that they should get themselves arrested if necessary to prevent the auction.

In the end, the jocular county sheriff calmed the crowd and won some sympathy for himself.

"No, I don't like to do it," he said. "I don't like putting veterans out of their homes, either. But I've had to do it."

There were no oral bids for the farm. No one could have heard if there were. So the only bid of the day was a written offer given to the sheriff before the sale.

It was from the Land Bank Association of Willmar, which offered $75,598.08 for the ----------- farm. An attorney said the amount represents principal and interest due on the mortgage, plus the additional cost of the foreclosure sale.

ഇ൦ഇ൦ഇ൦

By the time the 1982–1987 farm depression ended, Land Bank Associations and their parent, Farm Credit Services of St. Paul, would become part of the largest farm debt restructuring in history. The St. Paul Farm Credit bank would report losses of more than $1 billion over a two–year span, and its recovery from the brink of bankruptcy would be remembered for a few years as the largest recovery by a financial institution in American history – a distinction that has now passed on to larger commercial banks. Still, personal tragedies won't be forgotten and will become part of the folklore of the countryside [3–11], just as "penny auctions" and farm protests are remembered from the 1930s.

One painful memory is lasting from my chronicling the farm depression of the 1980s. There were thousands of Midwestern farm families who resisted voluntary sales of land and foreclosures until all assets were lost. When they did leave the land, they left with nothing. Some left still owing debts that were not covered in mortgages or big, consolidated operating loans that became part of legal actions. Sure, some of these farmers had taken on too much debt; some were, as an insensitive former agriculture secretary insisted, poor farm managers. But most were civilian casualties of a war – a national war on inflation – that required abnormal sacrifices by producers of raw materials and made these farmers' land the primary battlefield.

ഇഇഇ

The pain and suffering was not restricted to farms and people working the land. It spread to communities and businesses that exist to support agriculture, by being marketing and supply centers for agriculture.

ഇഇഇ

Jan. 21, 1985 St. Paul <u>Pioneer Press</u> (headline:)
DEPRESSED RURAL ECONOMY HASTENS GHOST TOWNS
TRAIN STILL WHISTLES BY, BUT GEORGEVILLE IS DEAD

GEORGEVILLE (Minnesota) – The last indignity will be when the Minnesota Highway Department stops putting Georgeville on the map. That may be soon. There's no one left to be offended.

Georgeville has completed the metamorphous that threatens the assets of so many rural Minnesotans and the continued existence of dozens of rural towns, many of them much larger than Georgeville.

Its rotting buildings make a winter windbreak for several people who need inexpensive housing. Only a couple of its original homes remain occupied while a noisy, shaggy dog resides in a clapboard garage. Most of the area's residents live in trailer homes.

Tom Wold, an unemployed construction worker, has parked his trailer on what was once the parking lot for the Georgeville bar. The bar burned down six years ago. It was the last business in town.

"Not much has happened around here since the fire," said Wold, who made the 90–mile commute to and from St. Cloud last summer to work on construction projects. "There isn't anything to do right now. There's nobody left. It's over."

Wold pointed to another trailer where the husband and wife both work. "One commutes to St. Cloud every day and the other drives down to a job at Willmar (about 35 miles away)."

The trailers, of course, will be moved to more convenient lots if the Georgeville commuters get other jobs or gasoline prices increase.

Georgeville is west of Paynesville on Highway 55 between Regal and Belgrade. Regal and Belgrade are both in bad shape, too. Paynesville is still a strong farm service and tourism community, but its large Associated Milk Producers plant is operating at reduced capacity.

There aren't many jobs to be found in this part of central Minnesota.

Still, there are signs of life in Georgeville. Rabbit tracks break the snow in front of the empty general store. Soo Line engineers still blow warning whistles when trains approach the Georgeville crossing.

The wood frame houses are sagging, twisting and falling, hastening the day when some farmer will dig out the rubble and trees and plant corn. That has already happened at Crow River, south of Paynesville.

Crow River's Mennonite Church is still there. Meeker County maintains a garage for snow removal equipment and road vehicles. But only one family continues to live there, and the town's only traffic sign warns, "Cattle Crossing."

Converting Georgeville into a farm field will be hard work. The brick chimney and concrete floor at the old creamery site will be a demolition problem, and the solid brick frame of the general store still stands.

People once had faith in Georgeville. The general store once served as a country shopping center. Upstairs, there were rooms for rent, and it housed a grocery, a land office, a beauty parlor and a corner gas station. At one time it had an office for a bank.

Residents remaining in the area remember the store in more recent times as a headquarters for a hippie colony that tried to revive Georgeville. The building still stands tall, but its pink and lavender facade reminds the natives nearby of those last occupants.

As the colony packed up and moved out of town, one of its members took spray paint in hand and left an epitaph on the store's north wall: "And the prophet said, 'Your old men shall dream dreams, and your young men shall see visions, and where there is no vision, the people perish.'" [3–12]

ℰℰℰ

When change is accommodated, there are usually people in community, regional and national leadership positions who do have vision. And there are usually business people around who are also prepared for change. Sometimes, what passes for a vision for the future is a yearning for the "good old days," if they ever existed; no matter how improbable those days could ever be again. I've been a witness to examples of all these human experiences.

ℰℰℰ

World leaders gathered in Rome in November, 1974, with the proclaimed mission of "ending world hunger within a decade." Delegates from most of the world's nations gathered for this extraordinary United Nations–sponsored World Food Conference. Americans came in such numbers it nearly qualified as an invasion.

Hunger groups, religious groups, farm groups, individuals responding to their own private sense of a calling came with their own plans to end hunger forever. Leaders of these groups quickly learned that they were guaranteed a media audience if they attacked their own government's inadequate response to world needs. American groups were especially vocal in this practice, which was no doubt inspired by the lingering distrust of government from the just–ended Watergate scandal at home.

Senators George McGovern of South Dakota and Bob Dole of Kansas arrived at the conference, sized up the chaos, and then quietly slipped away to visit military sites in Italy where they had served during World War II. Though they were representing the Senate Agriculture Committee, there was little they could do. The United States' position on food aid programs and levels of assistance wasn't publicly known, and these two powerful senators didn't know it, either. Agriculture Secretary Earl Butz and Secretary of State Henry Kissinger were waging a power struggle within the new Ford administration over whose department was in charge of such international matters. The circus in Rome continued for several days, with Secretary Butz uttering some message nearly every day that nothing needed to be done to end food distribution problems after three years of a global drought cycle. When a European reporter asked him if there was enough food to go around, the secretary provided one of his classic, intemperate remarks. "Of course," he said. "Otherwise, they (the hungry) won't be here."

A few days later, Secretary Kissinger arrived at the conference with pledges of food aid to individual countries and to United Nations agencies. This raised new questions about who was representing the United States government, and whose department was in charge of U.S. food policy. And it also brought the private Americans from hunger groups out of the woodwork to denounce American aid as inadequate.

During this setting of bread and circuses, I met for breakfast one morning with Bill Taggart and Jim Webster, who were then agricultural aides to Senators Dole and McGovern. We shared cynical Washington laughter over the way Secretary Butz had infuriated Italians and Roman Catholics by defending U.S. birth control programs even before arriving in Rome. "He (the pope) no playa the game, he no maka the rules," Butz had said. And then, as we sipped our cappuccino, we talked about future public reactions to the food surpluses that were certain to come.

That morning, at a sidewalk cafe on the Via Veneto, I listened to the most intelligent discussion of the 16–day World Food Conference. Senate aides Taggart and Webster talked about the uneven line of progression one finds when tracking science. Hunger groups were worried that world population had crossed the "Malthusian curve," meaning that population had outstripped the world's ability to grow food and feed itself. But the Senate agricultural experts were convinced that science follows an uneven path, more like stairsteps. Gains in food production could cause a nearly vertical line increase in the amount of world food supplies. Then, production would level off, from restrictions imposed by weather or other influences, while it waited for the next scientific breakthrough. Both men knew of the work Dr. Norman Borlaug was doing to increase wheat yields in Mexico, and both knew of similar research to improve rice yields in the Philippines. Twenty years later, the uneven march of science has shown itself. Governments and agrarian policy makers have devoted far more time to coping with U.S., Canadian, European, Australian, Argentine and Brazilian surpluses in the past two decades than they have the economic and distribution problems involved with hunger. [3–13]

ഩഩഩ

Two months after the Rome conference, in January, 1975, I put down my copy of the English language Kuwaiti Times newspaper and my coffee. Men at the two tables near me in the Kuwait Hilton Hotel coffee shop were talking in Norwegian, a language of some prominence in my home area of western Minnesota.

After their shock that the eavesdroping America understood what they were talking about, if not all that they were saying, they told me their reason for being in Kuwait. Norway was opening its newly developed North Sea oil fields. The offshore oil reserves were enormous. And Norway knew it was about to reap enormous oil income – particularly if 1975 oil prices would survive pressure from the added supply coming on the world market. So, in a logical step to anticipate and prepare for change, a blue ribbon group of government officials, academics and business leaders was sent to Kuwait to study how the Kuwaitis were investing their oil wealth for the public good. "If we just make payments to our people, we would destroy our work ethic and Norway wouldn't be Norway," a team member explained.

The Kuwaitis, meanwhile, were investing in diversified portfolios of property and businesses around the world, preserving wealth and sources of revenue for future days when oil reserves are depleted. The Norwegians were rather impressed with the holdings of the Kuwait Investment Fund. They ranged from owning popular vacation properties in the Southeastern U.S. states to major equity positions in European auto companies.

ഇ൏൏ഇ

In the mid–1980s, Taiwan sent buying teams through the Middle West to visit state officials and agricultural leaders while reviewing and accepting bids on various grain orders, called "tenders." Most of these buying missions were public relations ploys. The grain trade has its own time–tested methods for receiving offers and making export bids. But the process did bring Taiwanese and Americans together to talk business and develop friendships.

On one such mission to the Midwest, I attended a grain buying session in the Governor's Office in the Minnesota State Capitol. Standing against a wall with members of the press were two German academics and a Danish government planner. One German was a linguist; he was visiting rural towns in the Midwest to discover nuances of the German language that were locked in time, before a century of technology and foreign influences teamed to change the German spoken in Germany. The other German and the Dane were in Minnesota to study the state's economic survival. The European Community was beginning negotiations for what was to become known as "the 1992 Plan," which would totally integrate Western Europe's economy by the end of 1992. How had Minnesota managed to keep its large number of headquarters companies and high–technology manufacturing plants when labor was cheaper and taxes lower in other states? Did Germany and Denmark need to fear corporate migrations to Portugal, Spain and Greece?

Their studies and reports probably helped sell the '92 Plan to Germans and Danes. Their initial findings after visiting Minnesota, Wisconsin and Washington state, they said, found a clear relationship between taxes and Northern infrastructure. If the Northern U.S. states sacrificed their education systems to lower taxes and thus "improve" their business climates, they probably would lose their companies. Had their findings been made widely available in the states, it may have helped Midwesterners and Northwesterners see the potential opportunities that may come from open borders and freer trade under the North American Free Trade Agreement and other trade pacts.

Several agricultural heartland states responded creatively to the agricultural crisis of the 1980s. Experimental finance and beginning farmer programs were started in states such as Iowa, Nebraska and North Dakota to mitigate problems with farm finance. Minnesota and Wisconsin started interest rate buydown programs, in which taxpayers picked up part of the interest costs on farm operating loans. And both Upper Midwestern states launched programs and study commissions to find ways to support their agricultural strengths while looking for ways to further diversify their rural and agricultural economies.

Gov. Anthony Earl of Wisconsin reached out to business, academic and agricultural sectors in 1985 to form an impromptu Wisconsin Department of Economic Defense to carry out the studies and make recommendations for state policies. [3–14]

Howard Richards, a farmer from Lodi, Wisconsin, who would later be named the state's Agriculture secretary, and Gary Rohde, dean of agriculture at the University of Wisconsin – River Falls, were named to head that unusual emergency department of government. Wisconsin had survived the farm depression better than neighboring grain producing states, Dean Rohde explained, because the federal dairy program gave Wisconsin's huge dairy industry better income protection. [3–15] But, he warned, dairy's turn for hard times was coming. "We have state and federal programs bumping heads right now," he said. "The new (1985) farm bill takes the old idea that we have too much milk, therefore we have too many farmers. Minnesota and Wisconsin are saying, in effect, that production is the problem, not the number of farmers, so they are trying to do their best in keeping their farmers in place during the tough times."

Looking ahead to state planning tasks, Dean Gary Rohde was extremely prophetic: "We need two types of programs. We need programs to help people weather the storm and we need programs to help people retrain and adjust to new lives when a lot of them will be leaving farming." Educators and farm organization leaders alike failed to see how long and deep the farm depression would run, he said. "As a result, the farm organizations and all of us probably spent too much time fighting to keep everything in place rather than working on the adjustment programs we also need." [3–16]

Agriculture did emerge from the farm depression, and farm income was strong nationwide from 1988 through 1992. But by then, both Minnesota and Wisconsin had separate dairy industry commissions in place searching for ways to strengthen the Upper Midwest dairy industry. And separately, dairy farmers in Minnesota, Wisconsin, Iowa and North Dakota were continuing legal battles in the federal courts, trying to force changes in federal dairy policies that set their milk prices as the benchmark, thus the lowest, for the country. Their state governments had joined the suit, and their stake in its outcome is significant.

ᎯᎯᎯᎯ

April 7, 1986   St. Paul <u>Pioneer Press</u> (headline;)

DIVERSITY EASES WAY FOR RIVER FALLS

RIVER FALLS (Wisconsin) – Val Lessard, manager of the Cenex feed mill, was talking about the 10 tons of dairy mixes he grinds each month when a woman came in and asked for rabbit feed.

"How much?" he asked.

"I don't know," she answered.

"How many rabbits do you have?

"One."

Lessard paused. His line of thought snapped. He would return later to discussing what big business the dairy industry is in western Wisconsin.

"Let's do this the easy way," he told the customer politely. "One rabbit. Ten pounds of pellets. A buck–fifty. Take this slip outside and the guys will load it for you."

In so many ways, River Falls is to western Wisconsin what Marshall is to western Minnesota. Both are college towns of about 10,000 people. Both are agricultural service centers fortunate to have faculty salaries and university support services in town to cushion the community's dependence on agriculture.

Both have significant nursing care industries partially tied to the aging farm sector. Elder people come here from farms and farm communities nearby when they need special care or want to be near medical facilities.

But comparisons break down on closer examination. Midway through the 1980s, River Falls finds itself a lot more fortunate. It is just beginning to show strain from the decapitalizing of agriculture that has occurred in this decade. So far, it has been spared the financial anguish so apparent in much of rural Minnesota, Iowa, Nebraska and other farm states.

The grain farmers, pork producers and cattle feeders in the other states watched the assets they built during the inflationary 1970s stripped away during the 1980s.

University of Minnesota economist Phil Raup has concluded that Minnesota's farm asset base had fallen back by mid–1985 to 1972 levels in constant dollars.

All that's left to show for the unprecedented rise in farm income and farm expectations from the 1970s is the farm debt acquired along the way to expand farms and purchase farm equipment at inflated prices.

Nothing comparable has happened here in Wisconsin. But it is starting.

The River Falls feed mill is a rare, corporately owned Cenex unit. The regional farm supply cooperative based in Inver Grove Heights (Minnesota) is a primary supplier of goods and services to locally owned cooperatives in Midwest and Northwest states.

Lessard has been a Cenex employee for only two years although he's worked here for several years. Cenex bought the River Falls facility when a privately owned elevator company folded the mill and elevator here and another at Hastings (Minnesota).

There was a second mill and elevator marketing grain and grinding feed for dairy cattle here until recently. But that one is now shut down.

Implement dealers are starting to fall by the wayside in Wisconsin, also. And at the University of Wisconsin – River Falls, Agriculture Dean Gary Rohde said area farmland values have now fallen by 26 percent – a substantial decline even if it is only half that reported by (Professor) Raup in Minnesota.

But (Dean) Rohde uses the figure cautiously. "There isn't much land changing hands," he said. "I don't know if you can determine the resale value of a dairy farm at this time."

Lessard agrees. "Land that is sold hereabouts usually becomes hobby farms for 3M Co. professionals and others who commute daily to the Twin Cities," he said. "There's a string of cars pulling out of here every morning, heading for the Cities."

And that's a big difference between River Falls and Marshall. River Falls is commuting distance from a major metropolitan area that offers employment opportunities when jobs are scarce close to home.

The gentrification of River Falls shows up on Cenex's books. "We sold twenty tons of bird seed last year," Lessard said. "You wouldn't do that in a more typical farm community."

Other differences are evident as well. Unlike farm supply businesses throughout much of rural America, the River Falls Cenex displays prices for products such as wood posts and barbed wire fencing materials.

These supplies are for farmers who continue to invest and improve their farm businesses. Cenex wouldn't sell them to farmers who are retreating from dairying and livestock.

But there is an even bigger difference between the River Falls feed mill and feed, seed and fertilizer companies in much of rural America. Dairy farmers can still get credit here in western Wisconsin.

"It's getting tight here, too," Lessard said. But so far, suppliers haven't switched to cash–and–carry feed sales like they have throughout most of the pork and beef areas of the Midwest.

Twice–a–month milk checks pay off the short–term credit extended by feed suppliers. But in a clear sign of the times, the co–op is requiring most of its farmers to have a line of credit from a bank or other lender on file backing up the credit it extends on feed and farm supplies.

"If things turn sour for a dairy farmer and we don't get paid some month, we can find ourselves with a $20,000 short–fall on the books," Lessard said. "It takes a lot of bird seed and 10–pound bags of rabbit pellets to recoup."

၈ဿ၈ဿ၈ဿ

Beirut, Lebanon, was a city filled with great Middle Eastern and European–style restaurants, and the rooftop of our hotel offered a combination of a great restaurant and nightclub. A rock group was playing modern Middle East music as we were taken to our table. But shortly after being seated, my wife and I broke up in laughter. The band unleashed a spirited rendition of "Bad, bad, Leroy Brown; baddest cat in the whole damn town..."

It was another reminder that Americans, and perhaps most of the world's inhabitants, are too willing to accept everything at face value. Arab musicians weren't supposed to know classic, inner–city music from the streets of Chicago, according to our convenient perception of things.

We have become conditioned, over the years, to know who the good guys were in World War II, and who were the bad guys. We knew the difference between the two groups in the Cold War that followed, as well. Our perceptions of the world were shaped by Secretaries of State such as John Foster Dulles, Henry Kissinger and Zbigniew Brezinski, to name just three, who were rather one–dimensional in their perceptions of the world. Their world, and thus ours, was based on power politics with only passing attention given to underlying cultural, geographic and economic influences that give shape to world events.

Travel, and getting to know people of different cultures, shatters myths – no matter how disruptive or inconvenient that can be to your orderly view of the world around you. Travel changes you; it conditions you to see and accept change. And it causes you personal pain when you see instances where change isn't quick or sweeping. I offer these travel experiences as examples:

ဢဢဢ

A Lebanese government employee we had met earlier visited my wife and me at our hotel one night in Beirut. It was early in 1975. The visitor correctly predicted that Beirut and all of Lebanon would soon be in flames. Its democratic constitution wasn't working, he said, because it only protected the rights of Lebanon's Christian, Moslem, Druse and small Jewish communities, with political power apportioned according to 1950s census data. A third, large, group of people was now living in Lebanon, and it had few constitutional protections and economic rights. These people, Palestinian refugees and Palestinian leaders in exile, weren't allowed to do much besides drive taxis, perform manual labor for wealthy Lebanese families, work for international agencies based in Beirut, and go to Lebanon's colleges and universities.

The economic strains this was causing Lebanon were fraying its governmental institutions. The whole fabric of this so–called "Switzerland of the Middle East" would soon be torn, though most Americans would not see this side of the Middle East from glimpsing Secretary Kissinger's shuttle diplomacy and his focus on power politics. The Lebanese government official knew his country was out of control. He asked my wife and me to sponsor him and his family so they could migrate to the United States before the civil war broke out. It was not something we could do – financially or emotionally. If we were to sponsor a family for purposes of U.S. immigration laws, it would have been one of the refugee families we had met in camps scattered about Lebanon, Jordan and Syria. We believed the Lebanese official had an obligation to his country and fellow citizens to stay and work at trying to prevent what was to come.

ဢဢဢ

A radio news reporter pushed a microphone in front of Senator Bob Dole's face and asked, "Senator, why do you want to be vice president?"

The senator didn't pause. "It's mostly inside work, and no heavy lifting," he said.

This was one of the memorable experiences that made 1976 a strange election year. Ridder Publications bought the Wichita Eagle and Beacon newspapers before merging with Knight Newspapers. Wichita's readers were especially interested in their Kansas senator's campaign for vice president that year, just as Minnesota readers of the St. Paul, Duluth and Grand Forks newspapers were especially interested in the vice presidential campaign of Senator Walter Mondale. Al Eisele, a colleague in our Washington bureau, traveled with the Minnesotan while I was assigned to Senator Dole. As good candidates for that office do, they both went where campaign managers directed and delivered prescribed campaign messages, even when the message didn't fit the locale.

Senator Dole was campaigning in San Antonio, Texas, one day when the message was particularly out of sync with the surroundings. He was delivering a rousing endorsement of President Ford's foreign and economic policies. The president had restored peace to the land, he said, "so we don't have all our young people in uniform." And, he added, as "good Texans" the audience should help the president commit America to the free enterprise system and "get government off our backs."

The large crowd gathered near the San Antonio riverfront was polite, but not enthusiastic. It was during the afternoon. San Antonio's free enterprisers were at work. At least 80 percent of the crowd were off–duty people from nearby military bases. Most were back from the collapse of our military operations in Vietnam, which was the cause for peace in the land, and nearly all were in uniform.

The absurdity of the scene would not be captured in "sound bite" reports of the campaign, either in print or broadcasts. Nor would a similarly absurd scene two days later at a sugar plantation near New Iberia, Louisiana. Senator Dole spoke in general terms about farm policy – the farm policy that affects the stereotypical family farms in the Middle West. There weren't any of those farm people present. Landed gentry from plantation families that pre–dated the American Civil War gathered close to the hayrack speaker's platform to applaud the candidate. Black field workers stood in the back of the plantation yard with the news media. Like the reporters, who never applaud or respond to political speeches, the field hands stood without emotion and watched the "white people's" political rally.

It was painful to watch. Nothing had changed since the Civil War except what has been imposed on this part of Louisiana, such as minimum wage laws, school desegregation and the right to vote. A way of life has changed, but not appreciably.

Around that same period, an outbreak of violence on the Pine Ridge Indian Reservation attracted national attention to western South Dakota. I was sent to the Pine Ridge to do a series of articles for newspapers that were then part of Ridder Publications. The series would explore different aspects of life on the reservation, but it would focus on the fight between modern, mostly halfbreed Christians who controlled power on the reservation, and the traditional, non–Christian and dual–religion people who followed the old ways.

It is painful to see people locked in place, denied the power and opportunity to succeed and pursue their dreams – if they still dream. I saw it happen to people living in Lebanon who didn't want to live there; I saw it among young military people who would soon be dumped out on a civilian economy, though the inflation and economy of the post–Vietnam War period discouraged entry into farming and business. It had happened before; Professor Ruttan's fellow soldiers had been replaced by tractors, as mentioned at the beginning of this chapter. I saw it in the Deep South, where manorial land ownership continues to have the unintended social and economic impact of segregation laws from an uglier period. And I was seeing it in my own backyard, on a South Dakota reservation where people had been put, not by choice, after their resources in the Black Hills were taken away. So isolated – so segregated – were the Oglala Sioux that they turned inward, on each other, trying to gain an upper hand in the tribe's politics and economics.

ଞଡଞଡ

Chief Frank Fools Crow, a great medicine man of the Oglala traditional people, visited me in my Washington office after testifying at Congressional hearings. During the visit, he told of the Dakota people's protest of a legal injustice that had occurred at Custer, South Dakota. The protest got out of hand. The Chamber of Commerce building was burning when someone suggested that the protesters should scatter to the wind. Several cars left in a caravan for Rapid City and the Black Hills. But enroute, they passed the highway entrance to the Chute Roosters resort, which has a covered wagon parked near a large arrow that points up into the mountains.

94

The covered wagon was too great a temptation for some of the Dakota people, Chief Fools Crow said. They stopped the cars and burned it to the ground.

The resort had insurance, he said. But the insurance claim – for a covered wagon destroyed by rampaging Indians – must have set off buzzers in Hartford, Connecticut, computers, he said. Investigators were sent to South Dakota to verify the claim.

Chief Fools Crow sat on a sofa in the late Walter Ridder' s personal office recalling the story. He broke into laughter, and used a motto from a New Jersey insurance company. "We Indians," he said. "We got a piece of the rock."

ዏዏዏ

A cafe on the main street of Martinsburg, West Virginia, serves the basic roast beef, roast pork or roast turkey sandwich platter with mashed potatoes and gravy. It also serves outstanding homemade pies. And the people who serve this fare, like many retail people in the mountain towns of West Virginia, are women from the nearby small farms who need the extra income to keep farming. The most memorable item in the restaurant, however, was the message from the proprietor on the bottom of the menu: "We hope your food and service was satisfactory. You know how hard it is to get good help."

Why hadn' t these hard working farm women become outraged and burned the place down? Perhaps working there was an example of the abuse people are willing take when their family farms are more a way of life than a business. Perhaps, also, it represents what is happening to people at a large number of American farm units that haven' t adjusted to change but are having change imposed on them.

And what has happened to those Ethiopian women and children, as described by Professor Raup, who were replaced by tractors?

ዏዏዏ

News reporters are always shown things and talked at. Our role is that of a professional bystander, or in the parlance of the countryside, a career hanger–on.

My role as such a person became certifiable in April, 1982, when I returned to Washington, D.C., to attend the annual meeting of the National Association of Agricultural Journalists. It followed a weekend at a Gettysburg, Pennsylvania, golf resort where Washington lobbyists, government workers and a few journalists who write about Western public policy issues gather for an occasional "North Dakota Open."

North Dakotans organize this combination unofficial golf tournament and visit to national battlefields. It was started by the late Scott Anderson and Washington attorney Tom Burgum, both of whom worked for Sen. Quentin Burdick, D–N.D., at one point in their careers; by sugar trade association executive Van Olson, who was once the administrative assistant to Rep. Mark Andrews, R–N.D.; and by the late Bill Wright, a long–time administrative assistant to Sen. Milton Young, R–N.D. But it is "open" to kindred souls who are concerned about life between the Mississippi River and the San Andreas Fault. On the Saturday night of the outing, an "awards banquet" is held at which time honorees are given a goose.

The award is actually the "Honker Hagen" trophy, named after a mythical North Dakotan. Legend has it that everyone has met this man, gone hunting or fishing with him, played cards or golf with him, and perhaps wasted an evening getting drunk with him. No reason for it; he was always there, the consummate hanger–on. I can't speak for former North Dakota Gov. Allen I. Olson, who also received a "Honker Hagen" that night, but I can see some parallels between many award recipients and this mythical North Dakotan – be he a Paul Bunyon, Pecos Bill or Forrest Gump. Myself included.

The goose sits proudly on my office desk at home.

Chapter 3 FOOTNOTES

1. I have long suspected, but cannot prove, that agricultural economists and the U.S. Extension Service have sparked a mild curiosity with deductive reasoning in rural America. Why else, for instance, would farm boys in my hometown ponder the following question (buried here in the footnotes for obvious reasons): "If a lamb is a ram, and a donkey is an ass, how can a ram in the ass be a goose?"

2. Agricultural economists at land grant universities were among the first and clearest explainers of the trade and monetary policy linkage. Therefore, some of the best economic analysis of America's economy in the early 1980s was published in Des Moines, Chicago, Kansas City, Wichita, Dallas, Houston, Twin Cities and smaller Midwest and Western newspapers, and not the large coastal newspapers, after an industrial recovery started in late 1982 and early 1983. One vivid example of this was the widespread regional attention given Harold Breimyer, an economist emeritus at the University of Missouri, who noted in 1983 that the Reagan administration abandoned supply–side economics and returned to Keynesian economics to combat the recession. This was done with increased defense spending and unprecedented deficit spending to "prime the (economic) pump."

3. Dr. Breimyer also explained to the Midwest why this government–generated business activity wasn't reaching into the region's resource–based economy. And since the recovery was closely linked to defense spending, some new business

was created for Silicon Valley companies. But Third World food customers strapped by global recession didn't immediately return to place orders for farm or technology goods.

4. Freivalds, John. Grain Trade: The Key to World Power and Human Survival. Stein and Day. Briarcliff Manor, N.Y. 1976.

5. From annual reports, Foreign Agricultural Trade of the United States (FATUS). Economic Research Service, U.S. Department of Agriculture.

6. Schuh, G. Edward. From speech delivered Oct. 6, 1984, to the Newspaper Farm Editors of America (now National Association of Agricultural Journalists) fall conference, St. Paul, Minn.

7. Robert Buckler, then a legislative aide to Sen. James Abourezk of South Dakota, was preparing a report on the Arab oil embargo for the Senate Interior Committee. He is now president, Issues Strategies Group, St. Paul, Minn. Clyde LaMotte was a syndicated energy columnist in the United States and Canada and owned energy news publications in Washington, D.C., before his retirement. He is also the author's father-in-law.

8. From a briefing by Nedloyd executives at the Port of Rotterdam in 1985 while on a sabbatical program sponsored by the Atlantic–Pacific Exchange Program.

9. Thierry Bechet visited Minnesota during a U.S. State Department tour program for foreign diplomats in 1993.

10. Names were removed to avoid embarrassment for people long after the events described.

11. For an example, see Holm, Bill. "A Circle of Pitchforks," in The Dead Get By With Everything. Milkweed Editions. Minneapolis, Minn. 1992.

12. From a series of articles in the St. Paul Pioneer Press, entitled "Troubled Farms," beginning Jan. 21, 1985.

13. For a detailed account of incidences and change in international food programs brought by the 1974 World Food Conference, see Ruttan, Vernon. Why Food Aid? Johns Hopkins University Press. Baltimore, Md. 1992.

14. From a series, "Agriculture's Flight of Capital." St. Paul Pioneer Press. April 7, 1986.

15. Ibid.

16. Ibid.

In October, 1987, I found myself using themes – if not facts – that G. Edward Schuh had been explaining to journalists willing to listen during his service in Minnesota and Washington. Participating in a Great Lakes shipping conference at Duluth, I couldn't help but notice the amusement shown by Canadian delegates whenever an American participant spoke of the Carter grain embargo and how it was still harming Great Lakes shipping volume.

Canadian grain shipments through ports at Thunder Bay, Ontario, and Vancouver, British Columbia, had dropped as much as exports shipped through New Orleans, Portland and Duluth. During my speech, I argued that Great Lakes shipping was an unintentional victim of changing monetary policies. The tight U.S. monetary policy raised the value of the U.S. dollar. Since most world raw materials are priced in U.S. dollars, U.S. monetary policy stifled trade for all raw material producers. The value of total global trade fell by 40 percent between 1981 and 1985.

Excerpts of that speech were published in Government Policies and Great Lakes Shipping: Perspectives on U.S. and Canadian Agricultural and Maritime Policies. April, 1988, University of Minnesota – Duluth. That account is further excerpted here:

"Most trade commodities are denominated in dollars. (World) wheat shipments fell from 101.3 million metric tons (mmt) in 1981 to 84.9 mmt in the 1985–86 shipping year. The U.S. share of that wheat market fell form 48.2 to 25 mmt. In coarse grains, where the U.S. has an even larger share of the market, the total global movements fell from 107.8 to 83.3 mmt, and the U.S. share fell from 70.7 to 36.4 mmt in four years.

"You cannot blame domestic farm programs and trade policies in Canada, the United States, Europe or Argentina. Every domestic public policy in the world didn't go bad simultaneously. And you cannot blame former President Jimmy Carter's Soviet grain embargo, although politicians on our side of the border have done so since 1980. The embargo is not the reason why Europe stopped buying spring wheat and durum shipped through the port of Duluth, or why Mexico stopped buying wheat from Canada, or why Nigeria stopped buying farm commodities from Brazil and Argentina. There have to be macroeconomic reasons for collapsing trade markets."

At the headquarters of Rabobank, at Utrecht, from left, former North Dakota Commissioner of Agriculture Myron Just, Berlin, N.D.; Co–Op Country Elevators President Dana Persson, Renville, Minn.; Lee Egerstrom, and Diana Carney and Harvest States Cooperatives board member Steve Carney, Peerless, Mont.

Photo courtesyof Allen Gerber

CHAPTER 4: THE NETHERLANDS EXPERIENCE

"If opportunity knocks, somebody ought to be home."
W. Brongers, spokesman, Campina Melkunie, 's Hertogenbosch,
Holland

All Northern European countries, the ancestral homelands of the vast majority of settlers in the Northwest, have modern–day cooperative businesses that can again inspire their distant sons and daughters in North America. Germany claims "the father of cooperatives." Great Britain lays claim to the first of the modern day cooperative businesses at Rochdale, England. Denmark has most of its agricultural production processed and marketed through cooperatives today, and its democratic form of government is an outgrowth of agrarian reforms and social justice struggles of 150 years ago that gave rise to farm cooperatives. [4–1] French cooperatives have achieved great success in new crops and technology transfer, finance and food marketing and processing.

But if you are looking for a complete, quick study of cooperative business opportunities, look to Holland and the provinces that comprise the Netherlands. Its agriculture is under stress from every conceivable force facing agriculture worldwide. Its dense and affluent population is pressuring available land for farming. It has a fragile ecosystem that requires care and some of the most rigid environmental restraints on agriculture found anywhere in the world. It has a highly productive agriculture that is losing its price and marketing protection from the Dutch and European Union (formerly known as European Community) farm programs. What remains of those programs are now endangered by provisions of the Uruguay Round of international trade agreements. And a very basic, fundamental problem of agriculture hangs over Dutch farmers' heads: They are not the low–cost producer of any of the world's foodstuffs, feedgrains, meats or oilseeds.

Despite such obstacles, Dutch farmers have one of the strongest agricultural systems in Europe. Their farmer–owned cooperatives, working in concert with government and academic institutions, are finding ways for farmers to survive changing world markets and farming restrictions. They have learned to work in strategic alliances with Europe's large multinational food companies that have worldwide marketing expertise they don't have in–house. They have learned where they can compete against the giants in the world's food industry, and they do, choosing their battlefields carefully. But they would rather change than fight. So they change, carving out niche markets they can serve and dominate, assuring a prosperous future for their farm members. Perhaps more than farmers in any other country, they are showing ways to position themselves for the Twenty–First Century.

കൗകൗകൗ

Few Dutch farmers have as bright a future, however, as do Mr. and Mrs. M.W. van Leeuwen at the southern Holland community of Delfgauw. In about 10 years time, they figure, the city of Delfgauw or a large industrial company will need their land for residential or industrial development. They know they will be gone from their centuries–old family farm by the time their pre–school son is college age.

In most respects, the Van Leeuwen family finds itself in the same position as American farmers living near the Twin Cities, Milwaukee, Des Moines, Portland, Denver, Sacramento, Orange County communities, Hartford, Dallas, Richmond, Charlotte, Atlanta or anywhere else where urban sprawl is gobbling up traditional agricultural land. But the Van Leeuwen's aren't simply waiting for the day their land becomes too valuable to farm.

"It's coming. We know that," Mr. van Leeuwen told visiting Americans during a December, 1992, visit to his farm. Even though they are preparing for the eventual, there is disappointment, he admitted. The Van Leeuwens know they won't be able to keep the farm operating until their retirement and won't turn it over to another generation of the family.

At prevailing land prices around Rotterdam area municipalities, the Van Leeuwens know they are sitting on an estate valued at more than $3 million U.S. dollars. This nestegg should give them the option of early retirement, said Mrs. van Leeuwen. They will also have the capital to move, buy another farm in a more remote area of the Netherlands, migrate to farms in other Western or Eastern European countries, or follow the route of some of her relatives and become successful dairy farmers in New Zealand.

With options available that include farming, Mr. van Leeuwen keeps improving the genetics in his herd of 70 Holstein cows, and he remains active in his dairy cooperative, Campina Melkunie. These two actions are inseparable for a farmer who wants to assure his family an appropriate farm income in the next few years and a long–term future in Dutch agriculture.

His herd is already twice the size of the average Dutch dairy herd, and he has quota rights under stringent national and European Union milk production policies that allow him to sell his herd's milk while limiting production on other farms. Improved genetics should allow him to keep the same level of milk production in future years while reducing his herd size and farm operating costs. At the same time, Campina Melkunie is aggressively trying to mitigate the consequences of reduced production quotas and falling European milk prices by raising the value of its 12,000 members' milk above European support prices. To do so, it is constantly researching and developing new products from milk components, it has opened sales offices around the world and now exports dairy products and milk derivatives to 130 countries, and has expanded abroad by owning and operating subsidiary companies in Belgium and the United States.

It isn't happenstance that Campina Melkunie is trying to stay out front of changes in world dairy markets. Cooperation among individual farmers is historical in the Netherlands. It was, and is, a logical response to Dutch geography which is the most pervasive influence on all things in the country.

"We never had a choice but to cooperate," explains Ewoud Pierhagen, director of International Affairs in the Ministry of Agriculture, Nature Management and Fisheries at The Hague. "Sixty percent of our land is below sea level. From the earliest days, farmers have cooperated to operate the canals and dikes. Everybody cooperates, or your neighbor gets flooded."

The Netherlands Today

The Netherlands has a land area the size of Massachusetts, Connecticut and Rhode Island combined [4–2] Measured another way, the country has a land area that is equal to about 55 percent of either Wisconsin or Iowa, the two smallest states in the Northwest region that is the focus of this book.

Despite a population of about 15 million and the presence of 17 municipalities with populations larger than 100,000 residents, 65 percent of the nation's land area is still devoted to agriculture. This requires an obvious frugal use of land and space in the Netherlands.

But the most unique feature of Dutch farmland is its origin. With more than 60 percent of the nation's land resources reclaimed bottomland from the North Sea, maintaining agriculture means operating an elaborate system of earthen dikes, canals, reservoirs and individually managed drainage ditches. This has built a national culture dependent on citizen cooperation, as Director Pierhagen observed. Indeed, it makes cooperation – both formal and informal – the first line of Dutch national defense.

Agriculture provides direct employment for slightly less than five percent of the Dutch workforce, compared with two percent in the United States. Related food, beverage and agribusiness services account for about one–quarter of the nation's workforce. [4–3] These employment numbers are similar to those in the United States, on a percentage basis, although Dutch agriculture's importance to the national economy seems to be better understood in the Netherlands. The food and agribusiness industry also provides about 29 percent of the Netherlands' total manufacturing revenues. It enjoys a favorable balance of trade in agricultural products, but trade figures are distorted by trade flows through the Port of Rotterdam, the world's busiest port.

Verle Lanier, a Billings, Montana native serving as Agricultural Counselor to the Benelux countries (Belgium, the Netherlands and Luxembourg) in the early 1990s, says from 30 to 40 percent of the U.S. exports to Rotterdam are transshipped up the Rhine River on barges to other European countries or reloaded and sent to sea to other destinations in Europe, Africa or Asia. About 80 percent of those exports, which are now approaching $2 billion annually, consist of grains, oilseeds and non–grain feed ingredients, Lanier said during a reception for visiting Americans at the American Embassy in The Hague.

But comparisons between imports and exports in any industrialized country are fuzzy, at best. If one looks at total food and agribusiness shipments, the Netherlands would have a substantial balance of trade surplus, as does the United States. Counselor Lanier, for instance, notes that Dutch farmers and their cooperatives recover about $1 billion annually from the Netherlands' import expenditures by shipping food and agricultural products to the United States. Most of this trade involves cut flowers, a worldwide trade–oriented horticultural industry dominated by Dutch grower cooperatives.

To appreciate Dutch farm and agribusiness success, despite the country's restrictive environmental policies and limited land resources, review these European statistics:

European Union Workforce in Agriculture–1990
(Percent of All Employment Engaged with Agriculture)

European Union Average	6.6 percent
Great Britain	2.2
Belgium	2.8
Luxembourg	3.3
Germany	3.4
Netherlands	4.6
Denmark	6.0
France	6.1
Italy	9.0
Spain	11.8
Ireland	15.0
Portugal	17.8
Greece	25.3

* The vast majority in all countries are self–employed farmers.
Source: Eurostat, European Union

ကကက

Net Value Added Per European Union Agricultural Worker
U.S. $$ per worker (1990) Europe Index

European Union Average	*$14,042.70*	*100.0*
Netherlands	$32,306.40	230.1
Belgium	$30,688.56	218.5
France	$21,047.04	149.9
Luxembourg	$20,370.42	145.1
Denmark	$20,088.18	143.1
Ireland	$18,084.78	128.8
Great Britain	$15,365.70	109.4
Italy	$13,533.66	6.4
Germany (excld former GDR)	$13,088.88	93.2
Spain	$12,118.68	86.3
Greece	$8,789.76	62.6
Portugal	$2,677.50	19.1

* The USDA uses different measures in its **State Financial Summary,** Economic Indicators of the Farm Sector, Economic Research Service. But if each U.S. farm unit averages about two farm workers, whether operator and employee, farm couple or parent and working child, the U.S. figure would compare with Great Britain's and Ireland's at $15 million and $13 million, respectively.

Dutch Self–Sufficiency
(Production as Percent of Domestic Consumption)
Source: Dutch Agriculture Ministry

Item	1970/71	1980/81	1990/91
Wheat	51%	60%	57%
Other cereals	21%	12%	13%
Potatoes	162%	222%	248%
Sugar	102%	161%	194%
Vegetables	193%	201%	227%
Fruit	66%	54%	31%
Butter	333%	355%	1,483%
Cheese	232%	225%	267%
Beef & Veal	124%	141%	154%
Pork	204%	240%	271%
Poultry	388%	296%	212%
Eggs	148%	300%	315%

DUTCH VALUE ADDED AGRIBUSINESS
Food, Drink & Tobacco Revenues
(Translated to 1993 $$ in billions)

Item	1980	1989	1990	1991
Total:	$30.7	$41.8	$41.7	$43.9
(of which):				
Dairy products	7.0	8.3	8.4	8.8
Meat products	4.5	7.0	7.5	7.8
Animal feed	5.1	5.6	5.1	5.1
Drinks/juices	2.4	3.3	3.0	3.2
Margarines, fats & oils	2.4	2.9	2.6	3.0
Cocoa, choc.& confect. sugar	1.5	1.9	1.9	2.0
Fruits & vegetables	1.0	1.5	1.5	1.8

* Statistics gathered from companies with 10 or more employees.

*

Auction & Wholesale Trade Revenues – 1990
(Translated to 1993 $$ in billions)

Area	No. Employees	Revenues
Ag commodities wholesaling*	50,000	$16.5
Food, drink & tobacco wholesaling	63,000	$32.5
Fruit & vegetable auctions**	3,900	$ 1.1
Flower auctions**	4,700	$ 2.1
Totals	121,600	$52.2

* Majority by cooperatives.
** 80–90 percent by cooperatives.

Source: Central Bureau of Statistics, Dutch Agriculture Ministry

ഇരുഇ

IMPOSED CHANGE

Change comes no less painfully in the Netherlands than in North America. The Dutch lost 30,000 farm units between 1980 and 1990, and now have slightly less than 100,000 farms. The United States lost about 300,000 farms during the same decade, but still has slightly more than 2,000,000 farms. The percentage loss was greater in the Netherlands. Not rounded off, Dutch farm units declined 23 percent while U.S. farm unit losses in the decade were only 12 percent, despite the financial hardship on American farms. [4–4].

Farmers on both sides of the Atlantic Ocean watched international trade negotiations between 1986 and 1993 with a mixture of excitement or trepidation, depending on the types of crops or livestock commodities they produced. They were justifiably concerned about what liberalized trade would mean to their markets and farm enterprises. The Uruguay Round of trade talks under the Geneva–based General Agreement on Tariffs and Trade raised worldwide awareness of agricultural and trade subsidies used by different countries and by Europe as a bloc. But even as the trade talks dragged on through the late 1980s and into the early 1990s, Europe began backing down from its high–cost subsidized agriculture system. As a member state of the European Union, the Netherlands imposed Europe–wide quotas on production and lowered prices on Dutch farmers while imposing its own farm and environment policies that restricted agriculture even more.

Price support protection and production limitations began declining under Europe's Common Agriculture Policy in 1985, a year before the GATT talks began. These reductions, or lowering of the "safety net," as farmers call it, continued on in subsequent years, much as price and income support began declining under U.S. farm policy with the passage of the 1985 farm bill. C. Ford Runge, an agricultural and applied economist at the University of Minnesota who served as part of the U.S. negotiating team in Geneva during the early part of the Uruguay Round, predicted that Europe's system of farm subsidies would "collapse under their own weight" with or without a trade agreement. [4–5] Members of the Dutch Parliament who work on farm policy issues were resigned to that fate in 1992. Members of Parliament Jan van Zijl of the Labor Party, P.M. Blauw of the Liberal Party, Pieter ter Veer, Democrats 1966 Party, and J. van Noord, of the leading Christian Democrats, agreed that the Dutch government and the European Union would accept a GATT that reduces and eliminates farm income maintenance programs. Early readings of the agreement, which was reached in December, 1993, appear they were right.

Most of the pressure for lowering price support policies has come from the burden agriculture places on the European Union's budget, said Mr. van Zijl. But some of the lower price adjustments are coming from European countries needing to accommodate other member nation's agriculture under the EC (European Community) '92 Common Market integration plan that went into effect on Jan. 1, 1993. That agreement attempts to create an open European market like that enjoyed by American states under the U.S. Constitution's Commerce Clause.

Regardless of future cutbacks, those already employed by European budget cutters were having an adverse impact on some farmers' income by the early 1990s. Dairy farmer Van Leeuwen, for instance, was receiving 80 cents (Guilders) for a liter of milk at year's end 1992, which was down from 87 cents in 1989. At currency translations then in place, and with metric conversions, Mr. van Leeuwen was receiving the equivalent of about $15.70 for 100 pounds of fresh milk. Under the United States' quaint federal dairy policies that ignore inventions such as modern transport and refrigeration, the price at Delfgauw was similar to farm milk prices in Florida, Texas, Arizona and California, and about $2 for 100 pounds more than the price paid farmers in Wisconsin, Minnesota and Iowa.

Though Mr. van Leeuwen's milk prices have been coming down steadily in recent years, his production costs are starting to decline as well. That's because most of his feed ingredients are purchased, except for his hay. The European Community was steadily lowering support prices for feedgrains that he used to feed his cows. And in late 1992, the United States and European Community were completing negotiations on an oilseed trade dispute. That agreement, he said, should lower his cost of buying soybean meal imported from North and South America.

The oilseed agreement prompted angry farm demonstrations in France, Belgium, Great Britain and in front of the Agriculture Ministry in The Hague during my visit there with Midwest cooperative leaders in 1992. Inside the Ministry headquarters, officials quietly conceded that the oilseed agreement was a net plus for Dutch agriculture even though it would harm feedgrain and oilseed producers. Livestock production was far more valuable to Dutch agriculture than the small amount of grain farming done on the country's precious land resources.

Away from the livestock and intensively farmed areas of the Netherlands, European grain farmers were making painful adjustments to changing farm policies. [4–6]. Matt Ridley, writing in The Economist, profiled the difficulty Great Britain's cereal grain farmers anticipated as they made plans to idle 15 percent of their land to comply with the European Union's 1993 farm program. By complying with the land set–aside scheme, farmers became eligible for government payments equal to about $145 an acre. But at the same time, farmers could expect a drop in cereal grain price supports that would lower the price of their grains by about $60 a metric ton. [4–7].

The predicament had a familiar ring to it. Though Europe's land set–aside program works differently than the traditional American supply–management schemes, the effect is the same: Lower total farm income. In April, 1993, Carl Schwensen of the National Association of Wheat Growers in Washington, D.C., was warning American farmers that economic projections of a proposed 1994 federal wheat program showed farmers losing both gross and net income by idling acres.

But it isn't just farm program changes that are removing Dutch farmland from production. Environmental policies, urban sprawl and development of recreation and "green" spaces are altering agriculture as profoundly, and may be more significant in future years.

The Dutch already operate the world's first "Manure Bank," which stockpiles surplus manures and seeks to find new products and uses for animal wastes. The bank also credits manure nutrients. In turn, the bank restricts farmers from applying more manure or commercial fertilizers to their fields than the amount of nitrogen and phosphate needed by the next growing season's crops. [4–8] Such environmental measures are plain enough to understand in the Netherlands. Groundwater resources come within a few feet of the land surface in many parts of the country, so surplus manure will leech nutrients into the water. Canals and ditches are likewise certain to carry runoff into water systems throughout the nation.

Gerard van der Grind, an economist with the General Policy Department of Landbouwschap, the umbrella organization for all Dutch farm groups, says current environmental laws demand that farmers reduce farm chemical use 50 percent by the year 1996 and cut use from 80 to 90 percent by the year 2000 on most field crops. Dutch horticulture, an extremely intensive form of agriculture, is farther along in curbing chemical use. Hans Zandvliet, public relations director for the Bloemen Veiling Flora flower auction cooperative at Rjinsburg, said greenhouse and garden farmers had already eliminated 90 percent of their traditional chemical use by 1990. Natural predators, such as certain types of ladybugs, are released in greenhouses and flower fields to control damaging pests.

Not all Dutch farmers recognize their country's environmental vulnerability or welcome restrictive public policies placed on farming. But Werner Buck and J. Schotanus of the Dutch Agriculture Ministry point out that success of existing programs hold out the only hope that new, more restrictive environmental policies won't be imposed in future years.

That isn't likely. Recent scientific studies in the Netherlands show gas emissions from stockpiled and surface–spread manures are contributing to acid rain problems that, in turn, inhibit plant growth. As a result, current policies call for the reduction of ammonia emissions by from 50 to 75 percent. How this can be done without reducing livestock herds and flock sizes isn't clear. Dutch farmers had better hope their Manure Bank soon finds new uses for animal wastes.

There can be little doubt that environmental actions will curtail more farming practices in the future and lead to still fewer farms. At the same time, farmland is rapidly being converted to other purposes to meet both recreational and environmental objectives. Gary DeCramer of the University of Minnesota's Humphrey Institute studied "water farms," for example, while he was a participant on an Atlantic–Pacific Exchange Program sabbatical in the Netherlands. Entire farms were being purchased by municipal and regional water systems to assure neither manure nor chemicals would be placed on the land directly above groundwater resources.

It seemed to be both a desperate and practical solution for protecting the integrity of municipal water supplies, Mr. DeCramer notes. But after returning to Minnesota, he learned that communities in the southwestern part of his state were creating their own water farms. Existing municipal water supplies were found to be contaminated with nitrates and farm chemicals. Deeper wells, reaching into lower aquifers, were being dug and the municipalities were taking steps to prevent their degradation.

"There are parallels and important differences with groundwater problems in Holland and in parts of the United States," he says. The greatest parallel, perhaps, is that agriculture must make environmental and business adjustments for water problems that are not exclusively of its own making. "The Dutch are quick to point out they have delta land. Poisons and contaminants come down stream at them from other countries along the Rhine (River). Regardless where it came from, the Dutch either deal with it or drink it," he said. [4–9]

Meanwhile, water problems in southwestern Minnesota and in vast areas of the Great Plains are equally difficult to stop at the source, Mr. DeCramer says. Carbon dating of groundwater resources shows some nitrates are thousands of years old. Some of it results form rock formations left in the soils by the advance and retreat of glaciers during the North American glacier age. But, Mr. DeCramer adds, it makes no difference how long the problem has existed. Modern agriculture cannot contribute more nitrates and chemicals to the water supply. The Dutch, with hundreds of years experience in engineering with water, are again serving as a laboratory for Americans and others around the world who are entrusted with preserving groundwater resources. One inevitable consequence is clear: There will be more restrictions placed on farming methods in the years to come.

Still more Dutch farmland is being diverted to state–owned nature reserves (Staatsnatuurmonument). Land held in these reserves increased from slightly more than 30,000 acres in 1980 to about 500,000 acres by 1991. [4–10] Protected wetlands increased from about 20,000 acres to more than 800,000 acres in that same period. Private conservancy organizations increased their land holdings in that time by 200 percent, to more than 125,000 acres. Separately, and for varied recreational and environmental reasons, the Dutch National Forest Service increased its land holdings by more than one–third, to more than 150,000 acres. [4–11]

Such takings are to be expected in a densely populated and affluent society. Affluence and population pressures will likely take even more land out of farming than the amount being diverted to the so–called "green spaces." I haven't learned how much land is being converted to such uses annually in the Netherlands. American experience suggests the conversion would have to be large. In a Nov. 30, 1992, publication,

American Farmland Trust noted that an average of two million acres of U.S. farmland is lost to urban sprawl annually. That average rate of conversion has continued since 1970.

"The outcome (of Dutch and European Union policies) will be a lot less room for everybody to stay in agriculture," concedes Mr. van der Grind at Landbouwschap.

"Isn't that a pretty negative assessment of Dutch agriculture's future?" asked Dana Persson, a visiting cooperative manager from Minnesota. The question didn't bring an immediate response. Finally, Mr. van der Grind answered with another question. "You say our message isn't positive, but what is the alternative?"

Those questions confront American farm organizations, agribusinesses and rural community leaders. The Dutch have a system for seeking answers, even when they can't assure the future existence of all their farms and farmers. It involves cooperation among institutions that Americans haven't matched, lacking both the legal standing and political will to do so.

This Dutch cooperation brings agriculture, government and academia together in the Netherlands in much the same way government and industry work together in Japan. Landbouwschap, for instance, is the umbrella body for all farm and production groups (farm trade associations) in the Netherlands. In turn, the Dutch government gives it quasi-governmental authority to administer parts of the national farm programs and work with coordinating agricultural research.

Landbouwschap also works with the National Cooperative Board for Agriculture, another quasi-governmental body made up of all farmer-owned cooperatives. And it works with the Agricultural Products Board, which brings in the privately-owned and public stock food and agribusiness companies to forge consensus opinions on national farm and food policies. Together, the collaborative players in the farm and food industries work with their own research departments, the Dutch Agriculture Ministry, the country's research universities and the national Extension Service in search of technologies and policies to keep ahead of world market changes. This collaboration is written into law, and by common practice – like English common law – that has taken on legal standing over time.

"Extension, education and research are the cornerstones of our agriculture policies," says Mr. van der Grind. "We have to help our farmers remain competitive in the marketplace."

A dean of agriculture at a land grant university, a president of a large farmer-owned cooperative or a farm organization leader in the United States might say the same thing about American agriculture and its similar institutions. But it isn't the same. When you visit the Netherlands and visit its companies and institutions, you see economies of scale that come to effective public policy and research that we, in America, see at state levels but rarely see duplicated nationally across the diverse interests, climates and cultural differences of the regions. Food and agriculture aren't bound as tightly together in America. The American food and agriculture industry's large companies tend to be centered in the five states of New York, California, Illinois, Minnesota and Missouri, based on business directories' listings of company headquarters. The extent that they work with land grant and public research institutions is shaped by how multinational they have become, how competitive they are with other companies and by need to protect proprietary trade secrets and technology. Moreover, the conflicting public policy signals from U.S. farm organizations offer evidence of our incohesiveness. So does the competition for public and private research dollars at research universities and Experiment Station research sites.

In contrast, much of the current research now underway at Wageningen University, the major agricultural research center in the Netherlands, is focused on finding alternative crops for farmers and more diverse value–added products for agribusiness to process or manufacture. There are similar, smaller projects undertaken at American land grant universities, but they do not enjoy widespread national support from the public or government. [See Chapter Note A] The economic benefits in finding new, less chemical–dependent methods of farming, or finding new, alternative crops aren't as apparent in the larger United States. But these benefits can be seen and understood in the smaller, more homogeneous Holland. "In general, we are trying to find more market–oriented products. People (farmers and their cooperatives) must produce more market–driven crops and less government subsidized crops," said Mr. van der Grind. "Prices are coming down!"

The Dutch Cooperatives

The following pages offer brief profiles of major Dutch cooperatives that are leading change in the Netherlands by increasing the value of their members' products. [4–12]

৪৩৪৩৪৩

Cebeco Handelsraad – Owned by half the farmers in the Netherlands through 40 local cooperatives, Cebeco has become for

Dutch farmers what directors and managers of Farmland Industries and Land O'Lakes envisioned when they discussed merging in the mid–1980s.

It has become, though on a smaller Dutch scale, the equivalent of a giant diversified inter–regional food and agribusiness cooperative. It had 1991 sales of five billion Guilders, or approximately $2.3 billion based on currency translations at the time of its annual report. About three billion Guilders came from its farm supply business while food product sales produced another billion Guilders and the final billion came from its agricultural processing, commodity handling and marketing businesses. "Our farmers need the supply business, but there isn't much money (profit) in it," explained I.J. Prins, the company's president and chairman of the executive board. "We give the farmer a good product at the lowest possible price. In principle, it's a low margin business. And the market is coming down," he added in reference to environmental restrictions on agriculture.

Growth, then, is coming from its international involvement with handling and marketing, and even more from its diverse expansions into food product processing and marketing. "Membership is nice, (but) money works," Prins said in explaining the movement into more lucrative food processing. So Cebeco has expanded by buying Dutch food companies in recent years, by acquiring processing plants and food companies in other European countries, and by taking a 51 percent stake in a Canadian prepared foods company. It's seed division also owns a seed and flower company at Halsey, Oregon.

President Prins said some of Cebeco's (Cebeco Handelsraad – Dutch cooperative) international operations were entered primarily for information and learning experiences. They allow the cooperative to "keep track" of world market developments and technologies, he said. For example, Cebeco is a member of the A.C. Toepfer grain marketing joint venture in Hamburg, Germany, which allows it to interact with its American partner, Harvest States Cooperatives. Prins said he also tries to keep current on other market trends through friendships at Land O'Lakes. As a partner in a Dutch egg auction with other co–ops, he added, Cebeco is cautiously watching Minnesota–based Michael Foods introduce new egg products and technologies before his company invests in egg processing.

By focusing more on processing and food products, Prins sees Cebeco as positioning itself to help farmers raise the value of their production. For instance, the co–op is already a major supplier, processor and marketer for the garden farms that grow fruits and vegetables that are largely unprotected by European Community agricultural programs. While this strategy is producing profits for

Cebeco's current members, it is more of a positioning the cooperative for the changing world food market. "Where is income for farmers coming from in the future?" he asks rhetorically. "The market."

Bloemen Veiling Flora – The seven flower auction cooperatives in the Netherlands offer a perfect example of flower farmers working together to create a world class niche market. Bloemen Veiling, at Rijnsburg near The Hague, is the third largest auction co–op in the country. It sells more than 5.8 million flats and bunches of flowers daily, which gives it a 10 percent share of the Dutch market.

The cooperative provides auction marketing services for its 1,572 member–owners and 567 other non–member growers in the Netherlands, says Hans Zandvliet, the auction company's public relations director. And to insure an adequate daily supply of cut flowers, bedding and potted plants the year around, Bloemen Veiling also provides auction services under contracts with 764 growers in Israel and 586 farmers in other countries. Because it has such adequate supplies and varieties of horticultural products, the Rijnsburg auction attracts from 600 to 1,100 buyers daily who buy for clients around the world. This, in turn, helps Dutch farmers and their seven cooperative auction houses control 70 percent of the world flower trade.

There are actually two markets in one at Bloemen Veiling, explained the public relations director. Most of the flowers are bought by buyers for large commercial accounts. They, in turn, ship fresh flowers to customers easily served by KLM and Northwest Airlines, and other European and Asian cargo carriers. Within 36 hours, says Mr. Zandvliet, these fresh flowers can be in retail shops at Omaha, Nebraska; Denver, Colorado; Sioux Falls, South Dakota; Fargo, North Dakota, and points throughout the eastern United States. The West Coast tends to be served by domestic and South American flower companies. But a third of Bloemen Veiling's flowers are bought and sold by the marketers who made Holland synonymous with flowers centuries ago. They are "the Flying Dutchmen," the wholesalers and truckers who buy flowers early, load their trucks, then go flying off across Northern and Central Europe's highways to supply retail customers on their routes.

Auctions are the same the world over, it should be noted. I've watched camels being bought and sold as pets by wealthy business people at an auction outside Doha, Qatar, and I've watched prospective race horses being auctioned to America's wealthy at a stable near Lexington, Kentucky. It's the same process used at cattle and hog auctions at Sioux Falls, Omaha, St. Joseph, Missouri, and Lancaster, Pennsylvania. Not much is different about auctions except the nature of the items being sold, and the technology and efficiency used in auctions.

The flower auctions are a success in the Netherlands because crafty people have found ways to use technology to make it the fastest, most efficient auction system in the world. For contrast, board a train at The Hague and visit the diamond exchanges at Antwerp, two hours away in northern Belgium. There, buyers and sellers gather to examine diamonds at the city's four exchanges that make the largest diamond market in the world. There is no new efficiency for diamond trading to be found in technology. Skilled buyers, who are artisans of the diamond trade, hold stones against the indirect light of huge north windows on the trading floors, continuing a trading practice that is believed to have started in Antwerp in the year 1447. [4–13]

The indirect light reveals flaws in the stone and other properties that will make the diamond easy or difficult for artful diamond cutters to turn into jewelry. On a visit to the Bourse Vrije Diamanthandel in 1986, Mrs. Marleen Beerens of the Diamond High Council regulatory agency spoke at length about the importance of northern light. She knew of only one other exchange in the world that was similarly structured, she said, and that was for "special grain buyers in America."

That exchange is the cash market for Northern grown crops at the Minneapolis Grain Exchange, where barley buyers check pigmentation against the northern light to find appropriate malting barley supplies for malt companies and brewers.

No faster, better way has been found to auction the world's diamond supplies. But each year less barley, durum and other grains are being traded on the cash market in Minneapolis as grain companies are able to link millers and brewers to their supply needs at country elevators. The Dutch flower cooperatives worry their flower market may have more in common with North American special grains than the diamond market at Antwerp.

Colombia, the South American country, has improved its horticulture and now has 11 percent of the world flower trade, said Mr. Zandvliet. Multinational trading companies are starting to compete for market share, bypassing auctions with contract growers in South America, southern Europe and Africa. This changing world market is causing the Dutch flower and greenhouse farmers to reassess their cooperative structures, he added.

The Rijnsburg cooperative's growers are considering a change in their bylaws to allow full membership to Israeli and southern Europe's farmers who use Bloemen Veiling's auction services. Management believes opening the membership will strengthen grower loyalties that in turn will ward off direct contracting with trading companies. A proposal to open the membership was defeated at the cooperative's 1992 general meeting, Mr. Zandvliet said. But in doing so, the membership instructed management to continue studying the proposal.

ഇ൝ഇ൝

Campina Melkunie – The dairy cooperative, based at 's Hertogenbosch, was mentioned earlier in this chapter. It has revenues of about $3 billion annually that is maintained by expanding abroad and from value added processing while Dutch dairy production is capped by European and national policies.

With Europe's borders opening, the cooperative's various units have expanded throughout the European Union countries. It bought one of Belgium's largest dairy companies in 1991, which already had sales throughout the Community countries as well as in Eastern Europe. Among its other foreign holdings are DMV USA, at La Crosse, Wisconsin; DMV Ridgeview, at Whitehall, Wisconsin; and Deltown Specialties, at Fraser, New York. The DMV companies produce pharmaceutical preparations, whey derivatives, other dairy ingredients, Mozzarella cheese and industrial cheeses. Deltown is the world market leader in making protein hydrolysate products for various industrial customers.

Given Holland's and Europe's dairy surpluses, the cooperative is having difficulty raising its members' dairy payments above Europe's Common Agriculture Policy support price, said W. Brongers, the co-op's public relations director. But under the circumstances of Europe's dairy industry, that shouldn't be construed as a failure by this Dutch co-op. C. de Wit, a director of the cooperative, said Campina Melkunie's success in finding new markets and developing new products has lifted pressure on Europe's policy makers to further restrict Dutch dairy production. There can be little argument that he is right.

Given a stable Dutch population base of about 14.5 million people, and a flat domestic dairy market, Mr. Brongers said Campina Melkunie will continue looking abroad for expansion and export markets for dairy products and derivatives. "If opportunity knocks, somebody ought to be home," he said. Dutch dairy farmers are willing to be that somebody.

ഇ൝ഇ൝

Suiker Unie – International trade negotiators were into their sixth year of haggling over agricultural trade barriers in Geneva when a group of Americans visited the sugar co–op in late 1992.

There was disbelief among the co–op's leaders when the Americans explained they came to see what Suiker Unie was doing right, not to argue and complain about Europe's sugar policies. H.J. Wolters, president of Suiker Unie, probably needed the most convincing.

The cooperative was formed in 1966 with the merger of four major sugar co–ops. The sugar industry in industrialized countries has been under stress from low world market prices in most years since that merger. Trade barriers and production subsidies have been needed to keep the industry alive in most of these countries. Suiker Unie is no exception.

Mr. Wolter's views on the likely impact of liberalizing trade under a GATT agreement was not substantially different than the views expressed by American sugar beet growers and their sugar cooperatives. But that aside, Mr. Wolter's co–op has gone to greater lengths than American co–ops to expand its members' income opportunities.

Its sugar plants produce 600,000 metric tons of white sugar annually from about two–thirds of the Dutch beet harvest. The plants also produce about 200,000 tons of molasses that is mostly further processed into alcohol, about 300,000 tons of lime for use as field fertilizer, and about 250,000 tons of beet pulp for use as livestock feed. To create uses for its sugar, it has expanded into food processing by manufacturing snack foods, syrups and toppings.

Since sugar beet production is also capped under European farm policies, and sugar beet land must be rotated, Suiker Unie has expanded into fruit and vegetable marketing and processing. It operates two companies that process and market spices, herbs and aromatic products that are used by the European food industry, and it operates a commodities trading and brokerage business with offices at major exchanges around the world. And, for its members own farming needs, it has its own agronomy services and supply business.

Within the realm of the latter, Suiker Unie has an international seed business with 110 years of experience from predecessor companies. Among its U.S. subsidiaries are Payco (corn), Interstate (sunflowers) and L.L. Olds (various seeds) that are well known in the American Midwest and Pacific Northwest.

ෂෂෂ

Eemshaven Sugar Terminal C.V. – Eemshaven is a new port built near Delfzijl on the River Eems estuary that divides the Netherlands from Germany on the North Sea shipping lanes. The sugar terminal is the largest warehouse and bagging plant at the port. It is mentioned here because it is a joint venture of sugar and shipping interests. Suiker Unie is a one–third owner of the terminal.

Director B. Klerks, whose position is similar to that of manager or president of a co–op or joint venture, said the port was built in 1988 as a transshipment center. Still in its operational infancy, the sugar terminal handled 250,000 metric tons of cargo in 1992.

The port and terminal make a convenient shipping center for Suiker Unie, which is based at nearby Groningen, and for the co–op's plants that are also located in the northern Netherlands. The location is more than convenient, it is strategic. Eemshaven has become an important port for shipping European Union food aid to Russia and Eastern Europe, which creates additional business benefits and income for the cooperative and its members.

ಐ೮೦೮೦

Avebe – Like sugar, potatoes are a depressed commodity in Europe and throughout most of the world. Avebe is a starch processing cooperative, based at Veendam, that is finding ways to improve earnings for its members' industrial grade potatoes.

It has become an international co–op. It has 6,600 farmer–members, of whom a quarter live across the border in Germany. Together, they manufacture 1,200 pharmaceutical and industrial products from potato starch and potato components and byproducts. Its annual sales are approximately $700 million. And it has expanded operations to foreign subsidiaries that manufacture products in France, Germany, Sweden, Thailand and at Janesville, Ohio.

Public relations director W. Veendorp said only 16 percent of Avebe's Dutch potato products are used by industries in the Netherlands. About 70 percent of its production is exported to markets outside the European Union market.

Even more impressive than its marketing skills and product development is the economic importance Avebe has on its members' income. The European Common Agriculture Policy support price for industrial grade potatoes was lowered to 11 cents per kilogram in 1992, down from about 15 cents in the late 1980s, Mr. Veendorp said. After the GATT agreement, the co–op's economists anticipate the price will fall farther, to seven or eight cents per kilogram. (A kilogram is equal to 2.2 pounds.)

"There wouldn't be a potato grown in the Netherlands at that price, Mr. Veendorp said, although some farmers in northern France have land and production costs that would allow them to remain in the business. In any event, Avebe's members don't rely on Europe's support price. Though Dutch farmers have production costs of about 12 to 13 cents per kilogram, shared earnings from Avebe's value added processing and related ventures have restored member payments to about 15 cents for each kilogram of potatoes they deliver to their plants. Even after GATT, and assuming no new markets are opened, the grower payments would still average about 12 cents.

Avebe and its members aren't waiting for world potato markets to sort themselves out after new trade rules take effect. They are looking for more business opportunities abroad and new growth markets to enter. That still won't preserve all of Avebe's current membership, Mr. Veendorp said. Some members won't meet government environmental regulations for reducing chemical use, and others won't survive declining prices regardless of the co–op's value enhancements, he said. "You ask what we will do in the future in Holland? The answer is, 'Nothing.'"

If all works as currently planned, Avebe members will see more expatriated earnings from overseas operations bolstering their incomes in future years. In early 1993, Avebe sent a team of co–op officials, scientists, marketing specialists and farmers to explore possible joint venture business opportunities with farm cooperatives in Minnesota and North Dakota. Another team was exploring a possible joint venture business in Nebraska, and Avebe was searching to buy a small corn mill in the American Midwest in 1994 to start a waxy corn processing operation.

శీలిశీలిశీలి

Rabobank – Capital formation is a traditional problem in agrarian areas the world over, whether for agricultural or other rural economic development. This need also offers one of the strongest arguments for creating producer–owned and consumer–owned cooperatives.

Little wonder then that countries with strong cooperative business cultures and strong cooperative enterprises have also established strong cooperative credit and banking institutions. The Dutch and the French offer the best examples, having created international banking giants from co–op ownership.

Rabobank is the Dutch farmer–owned bank formed 90 years ago as a full–service banking institution. It has none of the business limitations that were placed on the U.S. Farm Credit System banks when they were chartered by Congress to recycle capital back to the countryside. Based at the city of Utrecht, Rabobank has grown to be the second largest bank in the Netherlands, and has emerged as one of the world's 50 so–called money center banks.

The bank has foreign affiliates throughout the developed world, including a large North American bank subsidiary in New York that competes with other commercial banks. And in recent years, it has had no hesitancy to compete with Farm Credit System banks and the system's two cooperative business banks, the CoBank at Denver and the St. Paul Bank for Cooperatives.

C.F. Broekhuyse, deputy general manager, said Rabobank's U.S. operations are primarily focused on being a lender to agricultural processing companies, regardless of whether they are privately owned or cooperative. Other bank clients include large agricultural production companies, such as the huge poultry farms based in the Southeast and South–Central U.S. states. "We have the pretension we understand their business," he said in a modest response to a visitor's question.

The bank is unique among cooperative financial institutions, owing to its unusual structure. A.M. Dierick, chief economist, noted that the bank can neither be sold nor taken over by other means. It is technically owned by 729 local Rabobank member institutions and their owner–borrowers, much like a federated inter–regional cooperative business in the United States. It doesn't distribute retained earnings to the locals. Rather, it uses retained earnings for investing in new ventures and expansions, and reduces its service costs when earnings pile up. Its farmer–borrowers benefit from discounted capital costs on loans that can be as much as two percent below prevailing market rates.

As a result, Deputy Manager Broekhuyse said Rabobank has nine out of 10 Dutch farm mortgages in its portfolio, and better than 30 percent of the nation's residential home mortgages as well. Given its dominance in agricultural finance, Rabobank will have a profound influence on the shape of Dutch agriculture in future years. And Rabobank doesn't hesitate to play the money trump card.

"The subsidy system built up in Europe, which is a little different than the United States', is just unacceptable for our taxpayers," explains Economist Dierick. "Contrary to (European Union) policy, reality says you should produce where it is most efficient to do so." In keeping with that philosophy, Rabobank officials said they will no longer provide loans for farmers to buy quota rights from other farmers to expand dairy herds or expand production of restricted crops. The bank does stand ready, they added, to help farmers diversify into new crops and to help their cooperative business expand value–added production of new products from new manufacturing plants.

෨෨෨

There is a wonderful neighborhood in Rotterdam known as Oude Haven, or "old harbor." Apartments and townhouses line the canals, sharing waterfront space with pubs, restaurants and small retail shops. It is a great place for a mixed neighborhood of senior citizens, young professionals and industrial workers to live, shop and play. It is also a comfortable place for a foreign visitor to relax after a day of meetings and interviews to ponder what he has seen and heard.

Have the bankers at Rabobank become hard–hearted with small farmers who want to continue farming the way their parents and grandparents farmed? Of course they have, and they will spare their borrowing customer–owners a lot of financial and emotional hardship in years to come when government policy or global economics dictate they can't produce more highly–subsidized, high–cost crops. How long will it be before American bankers refuse to write mortgages for small dairy and hog farms that require seven–day–a–week attention by a single family?

The waitress interrupts. But before she can return with a Dutch beer and a plate of hors d'ocuvres, a fresh question comes to mind.

Do American bankers and farm lenders realize they are already restricting mortgage money to small livestock producers? It has to be happening now, though no banking or lending group has acknowledged having such a policy governing farm mortgages. But every lender studies business plans, cash flow prospects, and projected returns on capital, equity and labor before writing the loan. Even the Farmers Home Administration, the acknowledged "lender of last resort" that tries to use its federal lending programs to help new people enter farming, will carefully look at lifestyles and returns on labor before the 1990s end.

Everyone needs vacations and weekend escape times. A young couple entering livestock farming will need to show lenders how they plan to share livestock responsibilities with other family members, neighbors on nearby farms, or business partners. The third option would be farming on a large enough scale to have employees – a practical solution. But this doesn't square with popular agrarian myths and attitudes about the size and structure of family farms in many U.S. states. And it necessarily restricts entrance to agriculture for most Americans.

Should Dutch cooperatives invest and expand beyond Holland? Of course they should, if their priority is to raise farm household income – not just farm prices – for their member–owners. How else can the same

123

number of farmers as live in Minnesota, operating farms on about half of Iowa's land resources, compete or share markets with European food companies such as Grand Metropolitan and Nestle and American multinational agribusinesses such as Cargill, Kraft General Foods, PepsiCo and Archer Daniels Midland?

Will American farmers and their cooperatives respond as quickly to the changing world marketplace as the Dutch? I was wrestling with that question when a family–operated barge boat chugged into the harbor and tied up at a pier in front of me at the sidewalk cafe. Its arrival was a welcome diversion. That's because I already knew the answer to my question. American farmers are not adjusting as quickly to changes in world markets and food systems. That was the reason for my going to the Netherlands.

Chapter 4 FOOTNOTES

1. Kirmmse, Daniel K. Kierkegaard in Golden Age Denmark. Volume I. Indiana University Press. 1990.
2. Briefing paper. American Embassy. The Hague, the Netherlands.
3. The Netherlands in Brief. Ministry of Foreign Affairs, The Hague. the Netherlands.
4. Dutch farm loss percentage taken from Eurostat. The European Community. Brussels, Belgium; and U.S. farm loss taken from Willard Cochrane, as cited in Anatomy of an American Agricultural Crisis. Rowman & Littlefield Publishers. Lanham, Maryland. 1992.
5. Runge, C. Ford. This was a point he made in several monographs and speeches presented in the late 1980s, and indeed, the European Community began scaling back farm income support in each subsequent year leading to the 1993 Uruguay Round agreement on trade.
6. For a concise description of Europe's farm programs, see A Common Agriculture Policy for the 1990s. Office for Official Publications of the European Communities. Luxembourg.
7. Ridley, Matt. The Economist. London, the United Kingdom. November, 1992.
8. From briefings at Landbouwschap headquarters and at the Dutch Agriculture Ministry, The Hague.
9. DeCramer, Gary. From a monograph report on his Atlantic–Pacific Exchange Program sabbatical, 1988.
10. Facts and Figures 1992. Ministry of Agriculture, Nature Management and Fisheries. The Hague, The Netherlands.
11. Ibid.
12. From interviews and annual reports, first assembled and published in the monograph: Egerstrom, Lee. Rediscovering Cooperation. Minnesota Association of Cooperatives. 1993.
13. Legrand, Jacques. Cut in Antwerp. I.U.M. Antwerp, Belgium. 1982; and from interview with Beerens, Marleen. Hoge Raad voor Diamant (Diamond High Council). Antwerp, Belgium.

124

Research is underway at most American land grant universities on alternative crops, "sustainable agriculture" and reduced chemical farming practices. However, expenditures for this research are small compared to support for livestock genetics and agronomic improvements for major, exportable crops. Clearly, research responds to public pressure and the availability of research dollars.

Notable work, akin to that underway at Wageningen University in the Netherlands, was launched in the early 1990s at Washington State University, Pullman, Washington; at various sites in Canada; Massey University at Palmerston North, New Zealand; and at Beltsville, Maryland, and other major research stations for the USDA Agriculture Research Service. For examples of breakthrough research that is not only environmentally friendly but economically sound, see "Sustainable Agriculture" by John P. Reganold, Robert I. Papendick and James F. Parr in Scientific American, June, 1990; and "Soil Quality and Financial Performance of Biodynamic and Conventional Farms in New Zealand" by John P. Reganold, Alan S. Palmer, James C. Lockhart and A. Neil Macgregor, Science magazine, April 16, 1993.

Professor Reganold, an associate professor of soil science at Washington State University, noted in personal correspondence of April 9, 1993, that the New Zealand study provided the first side–by–side comparison of the economic performance between conventional and biodynamic (like organic) farms. "The biodynamic farms in the study had no risks of chemical pesticides to human health and the larger environment and were just as financially viable as their neighboring conventional farms," he wrote.

Meanwhile, serious public advocacy and research groups across the nation that also recognize farm profitability to be part of sustainable agriculture had banded together in late 1992 to encourage the Clinton administration to change federal policies towards farm programs and research. Groups forming the Sustainable Agriculture Coalition included the Alternative Energy Resources Organization, Helena, Montana; Center for Rural Affairs, Walthill, Nebraska; Kansas Rural Center, Lawrence, Kansas; Land Stewardship Project, Marine on St. Croix, Minnesota; Michael Fields Agricultural Institute, East Troy, Wisconsin; Minnesota Food Association, St. Paul, Minnesota; Northern Plains Sustainable Agriculture Society, Wales, North Dakota; Organic Growers & Buyers Association, New Brighton, Minnesota; Sierra Club Agricultural Committee, Meadow Grove, Nebraska; and the Wisconsin Rural Development Center, Mt. Horeb, Wisconsin.

They advocated about 50 changes in public policies to enhance opportunities for lower input farmers and soil preservation measures. While they were still waiting a year later for official action on most of their recommendations, they were raising public awareness that federal policy can and does distort farming methods. These groups should be credited with seeing the need for public policy changes before contaminated water systems and other environmental problems force changes on agriculture, as is happening in the Netherlands.

CHAPTER 5 AMERICAN EXPERIENCES

I am looking over another old barn,
with the fading advertisement:
CHEW MAIL POUCH TOBACCO – TREAT
YOURSELF TO THE BEST. I say,
"Okay, I will," and get out my
package and make myself a good
chew. This barn is empty now –
empty of hay and straw, manure
smells, the assorted tools of
the farmer's trade. I recall
when the cowstalls were full,
but that was years ago, in a
better, happier time. The old
barn waits only for a painter to
paint it or a poet to put it
into a poem. But for the
present, no one loves it but the
prairie wind.

In Illinois, we have a Save Our
Barns Committee. They tell me
that they have saved a couple of
round barns out west of here.
Well, hallelujah and amen,
brothers and sisters! Who
knows, there might be something
left around here in ten years
that's worth looking at after all.

Dave Etter,. from "Barns," Selected Poems.
Spoon River Poetry Press. 1987.

The rural American landscape is littered with empty barns, candidates for creative advertising. Some serve as industrial size storage for bicycles, canoes and lawnmowers for the new rural residents who live on farms, but whose surrounding fields have been pried loose from the homesteads.

With few exceptions, American cooperatives aren't doing much to give these idled barns new reasons to live, to recapture the smells of animal agriculture or house the tools of the farmers' trade. This isn't a criticism of any U.S. incorporated cooperative. Rather, it is an acknowledgment that most American cooperatives are neutral to farm size. Some, on the other hand, find themselves functioning much like

American farm programs, rewarding growth, expansion and bigness both through volume sales of farm supplies and marketing larger volume farm commodities. Most, to their credit, are committed to serving member needs, large or small, commercial or hobby ventures.

This is logical for two undeniable reasons. First, American cooperatives face far greater competition today from private food and agribusiness companies than their European counterparts, and must compete against highly efficient suppliers and marketers. Second, America's farmer–owned cooperatives have nowhere near their European cousins' market strength. As a result, little if any strategic planning is given to finding new agribusiness ventures that would serve the sociological cause of changing economies of scale to keep people living and working on the land.

New cooperatives, such as production cooperatives built within communities, can do so. And new co–ops are again forming as farmers seek new ventures to break free from the cycle of buying out their neighbors' land, or selling their's to the neighbors, to keep expanding those fields of corn, soybeans, wheat, cotton and other major commodities. Ways to break the cycle will be dealt with later, in Chapter 9. In the meantime, it must be noted that there is a strong base of farmer–owned cooperative businesses on which to build and from which to find models.

The Dutch cooperatives discussed earlier, and other farmer–owned enterprises in Europe do have peers in North America. Large, successful cooperative businesses have been formed in Canada, Puerto Rico and Mexico, as well as in most regions of the United States. While some of these producer cooperatives have emerged at the top tiers of American food and agribusiness companies, none has achieved the market shares of their European counterparts except in certain fruits, dairy and farm supplies.

"We celebrate our cooperative successes, even though they don't look so big when you look at business lists," says Frank Blackburn, a veteran Midwest cooperative leader. Using 1992 revenue data assembled by the National Cooperative Bank in Washington, D.C., Mr. Blackburn makes startling comparisons. The eight largest agriculture and food–related cooperatives in 1992, not counting the food wholesaling cooperatives and insurance firms that are consumer co–ops, were Harvest States, St. Paul, Minnesota; Farmland Industries, Kansas City, Missouri; Agway, DeWitt, New York; Associated Milk Producers, San Antonio, Texas; Land O'Lakes, St. Paul; Countrymark Cooperative, Indianapolis, Indiana; Mid America Dairymen, Springfield, Missouri; and Cenex, St. Paul.

They had combined revenues of $21 billion. That same year, publicly–owned food company ConAgra Inc. at Omaha, Nebraska, ranked 18th on the Fortune 500 list with revenues of $21.2 billion. The Fortune Magazine ranked General Mills in 68th place and Ralston Purina as 69th, both with approximately $7.8 billion in 1992 sales; Borden was 74th with revenues of $7.1 billion, and Chiquita, the banana and fruit company, was 115th with revenues of $4.5 billion. Farmland Industries led agriculture cooperatives on Fortune's list that year in, but in a distant 145th place with revenues of $3.4 billion.

Though this doesn't distract from cooperative business successes, it does show co–ops for what they are. They remain a business ownership alternative in America. That they provide services for their member–owners is an added bonus. And they allow owner–members an opportunity to share in profits from additional rungs on the food or industrial ladder.

The added benefit of profit sharing from processing and adding value to a producer's raw materials should be clear to most Americans who pay even passive attention to business, but such cooperative business ownership doesn't hold the appeal to Americans that it does to Europeans. This is a point that Mr. Blackburn makes when speaking to farm and community groups. "When I start bragging about the big co–ops, I usually bring this (Fortune 500) list up to put things in perspective."

ഇന്ദ്രൻ

There may be lower expectations for farmer–owned cooperative businesses in America. If that is the case, it owes much to the American cooperative movement's founding and the reasons co–op businesses were formed.

Michael Cook, who holds the Robert D. Partridge chair at the Department of Agricultural Economics at the University of Missouri–Columbia, sees seven distinct types of cooperatives operating in America. They include the Farm Credit System lending institutions, cooperative rural utilities, bargaining co–ops, marketing co–ops, local supply/marketing co–ops, regional supply/marketing co–ops and "new generation" co–ops. [See Chapter Note A] Statutory and regulatory limits hold back some of these cooperatives – most notably Farm Credit banks and the utilities. Original missions and past experiences with less successful ventures may be discouraging cooperative members from turning their businesses loose to stray into new fields.

ഇന്ദ്രൻ

For many of America's largest regional cooperatives, growth has come through diversification. Early on at many of these farmer–owned companies, diversification meant little more than trying to sell their members both feed and seed. The cooperatives were, and continue to be, based on wholesaling goods and services to local, or federated cooperatives. [5-1] There were exceptions, such as the processing and marketing work of Sunkist in California and Ocean Spray in Massachusetts. Land O'Lakes and Farmland Industries branched out into new business lines and climbed to higher rungs on the food system ladder. But far–reaching diversification in the heartland cooperatives only began in the 1970s with rapid expansion into new areas of operations. Reasons given for those bold moves often cited growing competition in the food and agribusiness industries from conglomerates, with tobacco companies joining the competition by buying food companies in the 1980s. Throughout these periods of transition, the co–ops said they needed to position their companies for growth in more concentrated and competitive markets.

Growth came fast through diversification, then went into reverse just as quickly. The co–ops, like the private and public food and agribusiness companies, began pulling back from diversified ventures when they started reporting huge operating losses in the early and mid– 1980s.

The cooperatives have expanded far beyond supplying both feed and seed. This diversification has not greatly reduced their vulnerability to business cycles and extraneous influences on the broader farm economy. Reacting as many private and public stock companies did during the painful agricultural financial crisis between 1982 and 1987, cooperatives retreated to those businesses they operate best. Land O'Lakes stopped processing turkeys and sold its Spencer Beef division to Cargill in the late 1980s. Agway, operating out of New York and New England, scaled back its operations. Farmland Industries sold its Far–Mar–Co grain marketing assets in 1985, with Union Equity at Enid, Oklahoma, purchasing most of them. With most of its operations focused on handling, storing and exporting hard red winter wheat, Union Equity was unable to survive a combination of less income from government grain storage, lackluster exports and poor crops that hit the company in the late 1980s and early 1990s. The result was that Union Equity sold its assets to Farmland in 1992. Red ink is a great motivator for companies, whether private or public, cooperative or joint venture, large or small. Pillsbury, for instance, pulled out of the grain trading business about the same time that Farmland transferred its terminals and grain houses to Union Equity.

Even Cargill, the world's largest grain trading company, has stopped trying to cover the North American landscape with elevators and grain buying stations. [5-2]

The retreat, if that is what it can be properly called, ended in the late 1980s and early 1990s with the return of profitability to agriculture. But companies didn't go racing off to expand into new lines of business. Farm and food–related businesses began a rapid and radical reshaping of their structures. This has produced more concentration of companies in different niches of the industry, with companies buying assets from each other to gain market shares where they can be within the top three market leaders.

ജ്ജ്ജ്

Cargill sold a string of Montana and Wyoming grain elevators to General Mills,[5-3] narrowing its territories of operations. In turn, it expanded into more value–added processing of agricultural commodities, becoming more like publicly–traded ConAgra and Archer Daniels Midland. At some point around 1990, the company began issuing statements with "boiler plate" language describing itself as a processing company, though Garland West, Cargill vice president for public relations, doesn't recall exactly when the public emphasis on processing was started. But the shift towards processing was predictable, given ADM and ConAgra successes one and two steps up the food ladder.

General Mills has kept grain elevators in Montana and a small presence in flour milling, but otherwise it has become an international food company based on two niches – packaged foods that include its Betty Crocker foods and Big G branded cereals, and restaurants that include Red Lobster and The Olive Garden. Kraft General Foods, which had been welded together in the 1980s by Philip Morris Co., sold its Bird's Eye vegetable business to Dean Foods in 1993, with the former saying it wanted to focus on its dry and refrigerated food lines. [5-4] At about the same time, International Multifoods was selling its refrigerated and prepared foods division in pieces to other food company giants, explaining that it was focusing more on its foodservice businesses. Pillsbury has sold its grain handling business, most of its flour mills, and turned its remaining mills over to ADM to operate in a joint venture. [5-5]

So pronounced was the refocusing of America's food and agribusiness companies that securities analysts began wondering in early 1994 if ConAgra shouldn't begin thinning its business sectors. Bonnie Wittenburg, food industry analyst with Dain Bosworth

Incorporated in Minneapolis, noted that ConAgra was the second largest U.S. food company ($22 billion in 1993 sales), but remains the only major U.S. company to operate across the entire food chain. [5–6] With operations selling feed, fertilizers and chemicals, it nevertheless is best known as a food company with meat packing, poultry, frozen foods, dry consumer products and flour milling holdings. To manage this conglomeration, Mrs. Wittenburg noted that ConAgra had more that 60 "presidents" of food and business lines. [5–7]

The ink won't be dry on this book before other major changes occur in who owns or operates what agribusiness and food product assets in the rapidly changing 1990s. In one sense, ConAgra used a well–orchestrated acquisition strategy in the 1980s and early 1990s to expand and become more like Cargill. [5–8] And Cargill, with $50 billion in annual sales, was repositioning itself to be more like ADM. [5–9] Cargill's reining in of its grain handling and buying operations, as noted by Montana writer T.J. Gilles, would have been unimaginable in the roaring, inflationary 1970s when there seemed no limit to North American grain exporting potential. Even from that earlier perspective, the changes at Cargill would not seem too dramatic. For comparison, who would have predicted back then that Land O'Lakes' primary competitors in nationwide dairy foods markets would become food company properties of Philip Morris and R.J. Reynolds tobacco companies, with their Kraft and Borden acquisitions? [5–10]

ဢဢဢ

These changes in the food and agribusiness industries deserve a closer look. They shed light on both the strengths and voids in the American cooperative companies' ability to compete and carve out niches against such powerful, global competition. Even more importantly, they show how American cooperatives have evolved and cross–bred one or more functions of the original generations of cooperatives. [See Chapter Note A] And in doing so, a look at the current state of America's – usually regional co–ops – helps explain the need for new generation cooperatives that will be examined more deeply in Chapters 8 and 9.

ဢဢဢ

In the Northwest, Harvest States Cooperatives has emerged as the largest grain originating and initial trading company even though it remains in the second tier of trading companies worldwide. It pulled back from diversification in the farm crisis of the 1980s, selling most of its Great Plains lumber and building supply outlets as well as its interest in Froedtert Malt Co. But it grew to regional dominance in grain handling through linkages with local grain marketing associations and merger with North Pacific Grain Growers at Portland. This effort was further aided by rival companies pulling back from the territorial market. [5–11] It also strengthened its position in world trade by buying Land O'Lakes' equity position in A.C. Toepfer International grain marketing joint venture at Hamburg, Germany.

Well into the 1990s, originating grain sales remained an important part of Harvest States' grain business. That means it is the marketing house for local cooperative elevators scattered along rail lines that stretch from the Mississippi River to Puget Sound. Its customers for that grain are, more often than not, the grain trading giants of Cargill, Continental Grain and Bunge who have found export customers in need of grain, although Harvest States makes some export sales directly with overseas buyers.

Harvest States' founders in the former Farmers Union Grain Terminal Association talked much about improving farm prices by eliminating the "middleman." They had succeeded by the 1990s. They had become their own middleman, and the largest such entity in the Northwest.

Not all Northwest–grown grain is resold to the multinational grain companies. Harvest States' remaining interest in the malting business is in a joint venture with a private French agribusiness firm that has established North American headquarters at the Minneapolis Grain Exchange. It has a successful food processing unit, which packs private label food products for other food companies and under its own Holsum Foods label. Reaching out of its territorial backyard, the food unit has become the largest importer of olives into the United States. And Harvest States also has a successful oilseed crushing and oil refining unit, Honeymead, that competes with ConAgra, ADM, Cargill and large regional processing companies.

There are several business steps taken by Harvest States that strike resemblance to actions taken by their European colleagues in the Netherlands. One such unit that goes straight to the heart of raising the value of Northwest farmers' grains is Amber Milling. The milling company supplies semolina flour to pasta manufacturing companies. Semolina is the flour ground from durum wheat. So with its other milling and processing units, Harvest States has become an important value enhancer for grains grown in the Northwest states. Amber Milling enhances the value of durum grown in Minnesota, North Dakota and Montana. In mid–1994, Amber Milling announced that it was building a new flour mill in Wisconsin that will diversify the unit into processing and milling other Northern grown wheat crops, not just durum.

Harvest States' malting companies enhance the value of barley grown in those three Northern states, plus Wisconsin, South Dakota and Idaho; and Honeymead is enhancing the value of soybeans grown throughout Harvest States' marketing and service territory. Also valuable, but for fewer cents on the bushel, Harvest States' grain handling and trading services help find markets for the hard red spring wheat, a high protein crop grown in Minnesota, the Dakotas and Montana; and Pacific Northwest white wheat, a low protein wheat with special milling and baking properties used in making pastries and Oriental noodles.

The farmer–owners of Harvest States, and the local grain elevator associations that are federated members of the co–op, pay close attention to grain handling information at Harvest States' annual meetings, recalls Myron Just, a former board member from Berlin, North Dakota. [5–12] But the bulk of the cooperative's profits come from the processing businesses, not the grain trading for which the co–op is widely known. Mr. Just, who is also a former North Dakota Commissioner of Agriculture, says the processing profits will be the manna feeding the co–op through changing markets in the next century.

There are questions, however, about what roles Harvest States will play in those future markets. Will it expand its grain handling territory, perhaps through mergers or joint ventures with cooperatives or private companies in other regions? Will it expand on its already strong base in agricultural processing, either by manufacturing more products from its members' commodities or by further inroads in the food industry, such as its olive business? Will it look north, into the Prairie Provinces of Canada, striking a business alliance with any number of Canadian cooperatives that offer parallel grain marketing and feed milling services for their farmer–members?

Perhaps the biggest question of all is how can the cooperative make maximum use of its profitable food processing business units? Put another way, will it need to expand farther into the high cost and fiercely competitive business of making and placing branded, consumer–ready products on supermarket shelves? That question needs to be asked of Agway, Sunkist, Gold Kist, Land O'Lakes, Farmland Industries, Associated Milk Producers, Mid–America Dairymen and other regional cooperatives that also have success with certain branded products, but may be avoiding further diversification and are keeping themselves down on the food chain to avoid marketing costs. Perhaps, then, a more creative question should be put to large regional cooperatives: Is there reason to expand the role of inter–regional companies, such as Universal Cooperative, which is essentially a joint buying and manufacturing company for large regional co–ops, or for starting new inter–regionals, to take on the challenge of marketing members' new food or industrial products?

A promising but small step in that latter direction began in January, 1994, when Minn–Dak Farmers Cooperative, at Wahpeton, North Dakota; American Crystal Sugar Co., at Moorhead, Minnesota, and Southern Minnesota Beet Sugar Cooperative at Renville, Minnesota, formed United Sugars Corp. It is a joint venture company to market their annual production of processed sugars. That immediately gave the new marketing company about $700 million in annual sales. It hired a veteran food industry and sugar company executive, Robert Atwood, to serve as its president while establishing a headquarters in Edina, a Minneapolis suburb easily accessible to the nation's food industry sugar customers and to Minneapolis–St. Paul International Airport. United Sugars could prove to be a valuable model for cooperatives producing more diverse products. [5–13]

Even more promising, however, is the institutional cooperation among the three sugar beet cooperatives that led to United Sugars' formation. Southern Minnesota and Minn–Dak were already cooperating in sales before United Sugars by operating North Central Sugar Marketing, a joint venture also based in Minnetonka, another Minneapolis suburb, which in turn operated North Central Distributing at Chicago. They also participated in Midwest Agri–Commodities Co., a byproducts and molasses joint venture based at Corte Madera, California, near San Francisco. The two parent cooperatives rolled North Central's operations into United Sugars.

Clearly, the sugar cooperatives' managers and members have seen benefits of cooperating together short of full merger. Shortly after United Sugars was formed, the chairmen of Minn–Dak and American Crystal announced they had opened new business talks with corn farmers in the southern Red River Valley area of Minnesota and the two Dakotas. A new cooperative, Northern Corn Processors Inc., was being formed in early 1994 to produce high fructose corn syrup, sugar's rival sweetener. United Sugars will be the marketing agent for the fructose, and both Minn–Dak and American Crystal made commitments to invest with the corn growers in building the processing plant.

United Sugars President Atwood explained at the time that fructose and sugar sweeteners would "leverage" the marketing company with its food industry customers. "We'll be able to take care of all our customers' sweetener needs," he said. Moreover, he added during an interview, the marketing company would be positioned to sell and distribute any new products the member companies might start making.

That marketing capacity could prove extremely valuable down the road. Minn–Dak has a Delta Fibre Foods division that is developing food ingredient products from beet pulp. And Joe Famalette, president and chief executive at the larger American Crystal, says his company is committed to more diversification along the lines of the European farm cooperatives. American Crystal has been tremendously successful in keeping farm families on the land, unlike the constant sprawl and changes in economies of scale on nearby Red River Valley wheat farms. At the same time, the North American Free Trade Agreement, liberalized trade under the General Agreement on Tariffs and Trade, and changing, post–Cold War politics were making the economics of mere sugar production highly vulnerable in the United States. [See Chapter Note B] Mr. Famalette says the co–op will examine ways to diversify into other foods and commodities, including ways American Crystal farmers may use their land resources in crop rotations with sugar beet production.

In the Netherlands, officials of Suiker Unie said they questioned why U.S. sugar beet cooperatives only worked with sugar when most beets are grown in three year crop rotation schemes. The Dutch co–op provides agronomic services and crop marketing services for its members' other crops. American Crystal members and managers, and their partners in United Sugars, the corn fructose processing cooperative and joint venture company, are clearly repositioning business ventures along the more aggressive, European–Dutch model.

"I know the food industry from the brands down. Co–ops and farmers have known the food industry from production up. We need expertise from both ends of the food chain in co–op management today to survive the global market," Mr. Famalette said in explaining why American Crystal was attempting to become a European–style food company co–op. The company clearly has European roots for making such a transition, notes Mark Dillon, director of public relations. Aldrich Bloomquist, the sugar beet growers executive who teamed with prominent growers to form the cooperative in 1972, drew from the model provided by Suiker Unie in the Netherlands when forming American Crystal. And in the years since, it has been cooperating with the Danish Sugar Company cooperative, Danisco, on improving and marketing their Crystal and Maribo brands of seeds. [5–14]

Other large farmer–owned companies with experience on the higher rungs of the food industry ladder were also making bold moves again in the early 1990s. Farmland Industries, which went through a definite period of retreat in the mid and late 1980s to get its finances and operations back on firm footing, has become especially aggressive in adjusting to the radically changing world food market.

In late July, 1993, Farmland President Harry Cleberg announced that Stephen Dees would be opening an international business office in Mexico City later in the fall. Farmland Industrias S.A. de C.V., a subsidiary corporation, was organized and incorporated in Mexico to expand Farmland's substantial operations below the U.S. border. Mr. Dees, an executive vice president of Farmland, would also serve as president and general manager of the Mexican company.

"Mexico is a growing marketplace for Farmland," said company spokesman John Hendel. "Our exports to Mexico are growing each year." The North American Free Trade Agreement was certain to bring more opportunities, he added.

Farmland's export business was still less than five percent of the company's $3.4 billion in sales at the time the cooperative formed its Mexican subsidiary. But most of its exports were from food product sales, which were also Farmland's fastest growing business segment. In 1992, Mr. Hendel said, food sales had grown to account for 25 percent of total revenues. The petroleum business, for which Farmland was founded, still produced 28 percent of revenues. And fertilizer sales and agronomy services also edged food sales with 26 percent of revenues. Mr. Hendel said Farmland expected its revenue mix would soon change, with food sales emerging on top by 1995.

Though Farmland Industrias was the cooperative's first foreign subsidiary, it wasn't the company's first international venture. Farmland has business alliances with companies abroad, and has formed a joint venture with a Scandinavian fertilizer manufacturing firm. There is no doubt Farmland intends to expand horizontally beyond its base of operations in America's geographic midsection. "Our venture with an international office is critical to Farmland's strategic positioning in important foreign markets," President Cleberg said in announcing the Mexican subsidiary. [5–15]

Farmland has been growing vertically as well, prompting some controversy with populist elements in agriculture. It has helped farmers expand their hog breeding and feeding operations, while its meatpacking plants have expanded to make the cooperative one of the nation's largest pork processing companies. In many ways, Farmland resembles several of the large European cooperatives as it helps farmers with genetics, feed and placement of hogs on farms. [5–16] American dairy cooperatives have long provided similar farm–to–market services for its members, but broad–based Minnesota and South Dakota farm organizations, and smaller groups in other states, have

137

criticized Farmland for intruding on family–farm livestock production. Similar concerns were expressed by members of the Iowa Pork Producers Association, prompting a 1992 meeting between the pork commodity organization and cooperative officials to clear the air.

Questions raised by Farmland's current business ventures go directly to both ends of farm and food marketing. Can a large co–op or any other type of food company assure itself of markets and growth potential without expanding abroad? Can a large company with costly meatpacking and processing plants simply assume its loyal members will keep a steady, timely and uniform supply of hogs coming to market on a convenient slaughter schedule?

While Farmland was addressing those questions to meet its long term needs, Land O'Lakes was groping with similar questions in dairy country. Though a dairy–based cooperative will take different turns along the way, Land O'Lakes' officers are looking at horizontal and vertical paths for growth as well. And they, too, are building again in the 1990s after red ink and retrenchments in the 1980s.

Land O'Lakes was back on sound footings in the 1990s and earned $57 million on sales of $2.6 billion in 1992. That allowed the cooperative to distribute $48.5 million in patronage refunds to its Upper Midwest and Western members in 1993, and wisely, Land O'Lakes was busy making changes rather than coasting on its successes.

The businesses in which Land O'Lakes was gaining market share and posting profits included some of the country's most mature food– and farm–related industries, which offered less and less room for annual growth. [5–17] It was securely positioned as No. 1 in butter sales nationwide, but consumers were avidly seeking low–fat substitute spreads. It had become one of the nation's largest livestock feed milling companies, but livestock production was actually declining in its Upper Midwest and Pacific Northwest primary markets. [5–18] It was a successful partner, with Cenex, in a joint venture business that sells fertilizers and farm chemicals, but farmers continue to use these important production tools more sparingly. And environmental regulation of these farming tools increases each year, further restricting their use.

Despite these shrinking markets, better than two–thirds of Land O'Lakes corporate profits were coming from selling goods and services to its farmer–members, not from converting members' milk and farm commodities into value–added consumer food products that have made Land O'Lakes a highly recognized brand name in homes throughout America. Well aware of their slow–growth markets, company officials are making big changes. [5–19]

In mid–1993, Land O'Lakes was completing arrangements for its own Mexican presence. It was forming a joint venture company with a large Mexican food cooperative. That company would be charged with expanding Land O'Lakes exports of both food and farm products into the fast–growing Mexican economy.

Land O'Lakes was also completing the purchase of an Eastern dairy foods company to expand the co–op's dairy base of operations. Richard Fogg, former vice president for dairy foods, said Land O'Lakes was constantly looking for acquisitions to expand its food business when he headed the unit. The co–op had produced $1.4 billion of its $2.6 billion in 1992 revenues from food sales, and had set a goal of $2 billion in food sales by the year 2000. Acquisitions would play an important part in reaching that target, he said. But from a negative perspective, Mr. Fogg said the geographical expansion was also necessary to assure Land O'Lakes a continuing supply of milk. For a variety of reasons, including discriminatory federal milk marketing orders, milk production was declining across Wisconsin, Minnesota and North Dakota – Land O'Lakes' primary pasture.

A third area of important change is showing up in the dairy and freezer cases at supermarkets across the land. A steady stream of new dairy food products – such as Land O'Lakes' lines of Spread with Sweet Cream products, Light Butter and reduced–fat deli cheeses – are flowing to grocery shelves and food service warehouses and kitchens. With 120 food scientists and engineers on its staff, and armed with an annual dairy research budget of $10 million, Land O'Lakes boasted of having one of the largest research and development departments among American food companies.

International development work provides Land O'Lakes with a slower, more methodical approach to expanding markets. The co–op uses its own company experience, members' farming knowledge and hired expertise to work on agrarian economic development projects in about 40 countries around the globe. Land O'Lakes works as a subcontractor, providing training and development services under grants from the U.S. State Department's Agency for International Development and international development banks, said Martha Cashman, vice president for international development. When these projects succeed in building local farm and food economies, she said, Land O'Lakes develops friendships and business ties.

"We keep our international development work separate from our export business work, so we walk the people over to the right offices here if a project leads to developing a customer," Ms. Cashman said during an interview at her office. [5-20] Not all of these projects could be called unqualified successes. The international development banks and nations, such as the Soviet Union when it built the Aswan Dam in Egypt, all have checkered results. But Ms. Cashman is especially proud of projects that are succeeding in transforming the dairy industries of Pakistan and Nigeria into efficient, modern systems. The latter project literally walked in the door, she recalls.

In the late 1980s, a graduate student at the University of Minnesota showed up at Land O'Lakes headquarters and wanted to talk to someone who might have technical materials that could be used in improving Nigeria's dairy industry. That visit led to Nigeria sending several dairy farmers and milk marketing people to Land O'Lakes for training programs. International development agencies have since sent Land O'Lakes experts to Nigeria, and Nigerian dairy farmers have attended Land O'Lakes training seminars in Minnesota.

The cooperative's international development work started in 1981 under former Land O'Lakes Vice President LaVerne Freeh, an agricultural economist. It is now producing international business ties and export customers, says Duane Halverson, vice president for feed, seed, member services and international development. One of Ms. Cashman's projects, for instance, led to Mr. Halverson's feed group opening a large joint venture feed milling business in Poland. Other commercial ventures are being planned in former states of the Soviet Union and in Eastern Europe.

Jack Gherty, president and chief executive officer of the cooperative, said Land O'Lakes will continue on its growth paths using international market opportunities, new product introductions, and geographical expansion through acquisitions in the years ahead. He knows what faces the company if it doesn't keep marching along. Mr. Gherty became president in 1989, when Land O'Lakes was finishing its recovery from the farm depression of the 1980s, and that means he presided over the co-op's shedding of some unprofitable business units and staff layoffs.

"I've never thought being big was the same thing as being better. (But) we have to grow. I've been here through staff reductions in the past. I don't want to be here for another one," he said. [5–21]

New products and innovative ways to do business must come from employees on all rungs of the Land O'Lakes corporate ladder, Mr. Gherty said. Following the path of other industries and a few companies within the food industry, Land O'Lakes has adopted a less hierarchical approach to plant and headquarters management and decision making. In 1991, for instance, it started paying employees a reward for good performance similar to the performance bonuses often paid top executives at American corporations. About 4,400 employees received $450 payments, or rewards, in each of the first two years of the program. No distinction was made between employees who were and were not covered by union contracts, Mr. Gherty said. "I can't tell you how helpful this has been," he said. "This helped make everyone realize they are part of the team."

A team attitude isn't always easy to accomplish in a big company that has had several years of strong profits, or when change brings an element of instability to the workplace. The profits always prompt more conservative people to recall the adage, "If it ain't broke, don't fix it." And rapid changes cause other employees to seek security in their own little spheres of business and office units. Land O'Lakes has had rapid and sometimes startling changes. It bought Norris Creameries, a milk processing and distributing company in the Twin Cities market, then combined its own consumer milk products operations into Norris. It then spun Norris off into a public company, Country Lake Foods, in which Land O'Lakes retained controlling interest. Still later, it bought out the public shareholders and again made Country Lake Foods a wholly owned subsidiary. At the same time it was seeking the best structure for its milk products operations in the Upper Midwest, it placed its large agronomy operations in a joint venture company it formed with Cenex, the regional petroleum and farm supply cooperative based in the St. Paul suburb of Inver Grove Heights. More recently, Land O'Lakes placed its biotechnology research subsidiary, Procor Technologies, into a joint venture with private investors to form the St. Paul–based GalaGen Inc. That company was considering an initial public stock offering in 1994.

Despite all these moves, Land O'Lakes will remain focused on businesses that it knows best, insists Mr. Gherty. "You won't see us go off on a radical departure from our business interests," he said. It that strategy holds forth into the start of the next century, Land O'Lakes' future growth will be within current business lines. Land O'Lakes members who have diversified farms, and their neighbors, may need to create new co-ops to process and market their other commodities.

Dr. Robert Cropp, an agricultural economist at the University of Wisconsin at Madison and consultant to cooperatives, notes that Wisconsin and Minnesota were losing from two to three dairy farms a week during the early 1990s. That decline notwithstanding, "There is still room for growth in all of Land O'Lakes' business sectors, but the surest way is to reach out territorially and to raise up the foods," he said. "We know there are going to be fewer and fewer animals in Wisconsin and Minnesota, so this can't be the only place where the company does business."

Having raised a lot of questions along the way, Land O'Lakes seems to be finding a lot of its own answers. And some of these answers bear striking resemblance to the horizontal and vertical formulae being employed by the large, successful European cooperatives. Yes, a farmer–owned cooperative can work successfully in partnership with private and public companies at home, and not just in joint ventures abroad. Yes, a well–run company can still eke out profits in mature business markets, especially when cooperative services are combined with marketing. Yes, farmers will tolerate their companies' operating large research and development departments when the payback is never certain.

But will Land O'Lakes need to expand its food processing and marketing business beyond dairy foods to improve profits? Will its farm supply joint venture, known as Cenex/Land O'Lakes Ag Services, need to tread on other regional cooperatives' turf to expand its markets and assure future profits? With young and new farmers not wanting to go into the labor–intensive dairy business, or not finding financing necessary to do so, will Land O'Lakes need to help small producer cooperatives and limited partnerships get formed among dairy farm families to assure continuing supplies of milk near its plants and bases of operations?

ഇൻഇൻഇൻ

Cenex, the petroleum supply cooperative founded by Farmers Union organizations in the Northern Tier states along the lines of Farmland's original objectives, has evolved differently than most of America's large regional cooperatives. It has diversified greatly from its earliest years. But its diversification is closely tied to its base – its petroleum buisness. Its horizontal expansion has been into broader services areas while expanding vertically with manufacturing and services that are closely linked to the company's original mission.

The latter moves are largely what food companies would call line extensions. In the 1970s, for instance, Cenex moved boldly into securing oil reserves by buying rights to develop oil fields from major oil companies that were then more interested in the value–added enterprises of refining and marketing. In the 1980s, Cenex strengthened its own value–added refining and manufacturing capacity. And it began working with its local cooperative members to improve their marketing and supply systems.

By the 1990s, notes Cenex spokeswoman Lani Jordan, the company had firmly established itself among the second layer of petroleum companies operating within the United States and held leading market share positions in vast areas of the rural Upper Midwest and Great Plains. And, she says, Cenex had become one of the five fastest

142

growing operators of convenience stores in the nation by helping develop and then supplying more than 500 "C stores" at local Cenex petroleum cooperatives.

But there are questions facing the future for Cenex, too. It's formula for growth has worked well over the past three decades. There may be limits to how far this formula can go. Changing rural demographics, changing agronomic practices on farms and environmental restrictions on use of its products, and what appears to be an inevitable geographic collision with Farmland and other petroleum supply co–ops create problems that will need creative responses.

෨෨෨෨

Many, if not most, of the great questions facing America's cooperative businesses and the future economic vitality of rural America must be answered on America's farms, not in corporate board rooms. The questions must by asked of the farmers who own and support the major co–op businesses. Will they tolerate their companies becoming more a partner in their production, even when it means a better chance of sharing in value–added profits once animals and crops have left their farms? And looking farther ahead, will farmers tolerate their local and regional cooperatives setting aside a portion of retained earnings or sufficient operating capital to invest in risky ventures that may diversify their rural economies?

The memberships and managers of the different cooperatives will answer those questions in ways unique to their own businesses and markets. Clearly, some will conduct an assessment of their industry segments and opt to stay where they are, working on new, related products and building their existing market shares. Others will conclude that sweeping changes are in order, with changes aimed at generating more processing and marketing income for their members' raw materials. The type of commodities the farmers produce will also shape the answers.

Without a doubt, a key ingredient in cooperative planning will be the impact world markets are having on commodity prices for the farmer–members. American Crystal's Joe Famalette sees it this way:

"If you look at the brands backwards, you can see what is happening in the food industry and what's happening to farming. Farmers have to ask themselves if they want to just be farmers or if they want to be farmer–businesspeople.

"If they want to stay a farmer, they will be growing crops under contract for someone else. If they decide to be a farmer–

businessperson, they will be growing things for themselves (their cooperatives)." [5–22]

Chapter 5 FOOTNOTES

1. For a good discussion of the role of wholesaling, see the first six chapters of Fite, Gilbert C. Beyond the Fencerows: A History of Farmland Industries Inc. University of Missouri Press. Columbia, Missouri. 1978.

2. Gilles, T.J. Great Falls (Mont.) Tribune. 1991 news stories.

3. Ibid.

4. Chicago (Ill.) Tribune, Nov. 2, 1993; and corporate news announcements, Dean Foods and Kraft General Foods, Nov. 1, 1993.

5. From the Associated Press, Pillsbury and Archer Daniels Midland corporate announcements.

6. Wittenburg, Bonnie. Research Capsule. Dain Bosworth. Jan. 17, 1994.

7. Ibid.

8. Based on an interview, Jan. 28, 1994, with L. Craig Carver, securities analyst with John G. Kinnard & Co., who said ConAgra managed acquisitions better than most companies. They did not pay premium prices for acquired companies, and had made most of the assets profitable. But he conceded that ConAgra would be challenged to manage its diverse portfolio in future years.

9. The author's observation.

10. Based on an interview in July, 1993, with Richard Fogg, group vice president for dairy foods, Land O'Lakes. (Mr. Fogg left Land O'Lakes shortly after the interview and has become chief executive of Orval Kent Foods at Wheeling, Illinois. It was a refrigerated foods unit of Pet Inc., of St. Louis, that was purchased by Mr. Fogg and venture capital firms from Minneapolis and Milwaukee.)

11. Based on Harvest States Cooperatives corporate announcements and published articles in the St. Paul Pioneer Press, St. Paul, Minnesota, and AgWeek, Grand Forks, North Dakota.

12. From conversations with Myron Just, Berlin, North Dakota, at Rotterdam, the Netherlands, and St. Paul, Minnesota.

13. Cooperative business leaders point out other models do exist. Co–op Brands Inc. is one more prominent example.

14. Dillon, Mark, director of public relations, American Crystal Sugar From interview conducted in April, 1994.

15. From Farmland Industries corporate announcement, July 28, 1993.

16. Five cooperatives handle most of the genetics, feed formulation, slaughter, processing and marketing of hogs and pork products in Denmark, which is regarded as having the most sophisticated pork industry in the world. For a comprehensive description, see: Jurgensen, Erik Juul. Denmark: A Northern European Centre for Food and Agroindustrial Activities. Institute for Food Studies & Agroindustrial Industries, Copenhagen, Denmark.

17. A version of this material about Land O'Lakes first appeared in "The Search for Greener Pastures," St. Paul Pioneer Press. July 25, 1993.

18. California moved ahead of Wisconsin in dairy production in 1993, more from the decline in Wisconsin than the industry's growth in California. National Agricultural Statistics Service, U.S. Department of Agriculture. October, 1993.

19. Ibid.

20. See: "The Search for Greener Pastures," St. Paul Pioneer Press. July 25, 1993.

21. Ibid.

22. Egerstrom, Lee. "Keeping Things Sweet at Crystal Sugar." St. Paul Pioneer Press St. Paul, Minnesota. April 11, 1994.

Chapter 5 CHAPTER NOTE 5–A

Dr. Michael Cook, agricultural economist at the University of Missouri–Columbia, offers the following description of the seven business structures that have evolved for American agricultural–related cooperatives. It is excerpted from Food and Agricultural Marketing Issues for the 21st Century. The Food and Agricultural Marketing Consortium. University of Missouri. 1993.

1. **Farm Credit System.** Twelve Federal Land Banks were the first components of the Farm Credit System when it was chartered by Congress under the Federal Farm Loan Act of 1916. Subsequently the Federal Intermediate Credit Banks were created in 1923 to provide short– and intermediate–term credit; the Production Credit Associations in 1933; the Banks for Cooperatives in 1933; and the regulator, the Farm Credit Administration. The motivating forces behind the efforts to organize the system came from concerns about the unavailability of agricultural and real estate loans, extremely high rates and the length of terms (federal law prohibited national banks from making loans with maturities beyond five years.) After an initial surge of lending, the Farm Credit System loan volume continued to increase steadily until hitting a peak of more than $80 billion in outstanding loans during the early 1980s.

2. **Rural Utilities.** Formed to provide a missing service due to the high per unit cost of serving a low density customer base, the rural electric and telephone cooperatives were formed in 1936 and 1949. ... The resulting systems are a combination of approximately 1,200 cooperatives and 950 non–cooperatives receiving government subsidized loans providing telephone and electric service to more than 45 million rural customers.

3. **Bargaining Cooperatives.** (Sapiro I Cooperatives). Bargaining cooperatives address market failures through horizontal integration. Producers organize these Sapiro–inspired associations in an attempt to affect the terms of trade in favor of members when negotiating with first handlers. The functions of bargaining cooperatives can be described as twofold: (a) to enhance margins and (b) to guarantee a market (prevent post–contractual opportunistic behavior). These types of associations are found most often in perishable commodities in which temporal asset specificity creates a situation of potential post–contractual opportunism.

4. Marketing Cooperatives. (Sapiro II Cooperatives). Marketing cooperatives are a form of producer vertical integration pursuing a strategy of circumventing and competing with proprietary handlers. They usually can be categorized in one of two ways, single or multiple commodity. The objectives are similar – to bypass the investor–owned firm, enhance prices and, in general, pursue the Sapiro goals of increasing margin and avoiding market power. Because of property rights and benefit distribution issues, management and government functions are considered more complex in a multiple commodity marketing cooperative.

5. Local Associations. (Nourse I Cooperatives). Local cooperatives are economic units operating in geographical space where achieving scale economics in commodity assembly (usually grains or oilseeds) and input retailing might dictate the presence of a spatial monopolist/monopsonist. Founded to provide a missing service or to avoid monopoly power or to reduce risk or to achieve economies of scale, they epitomize the Nourse philosophy of cooperation –that of a "competitive yardstick" with the objective to keep investor–oriented firms competitive. Until the rapid expansion in the 1920s of regional structures, local associations were the predominant type of agricultural cooperative organization. Today, after much consolidation, local associations still are the largest type of cooperative in number.

6. Multi–functional Regional Cooperatives. (Nourse II Cooperatives). Competitive yardstick–driven regional cooperatives usually perform a combination of input procurement, service provision and/or product marketing. Many integrate forward or backward beyond the first handler or wholesaler levels. They might be organizationally structured as federated, centralized or a combination. They differ from Nourse I local cooperatives in that there is little probability of being a spatial monopolist/monopsonist in their geographic market. Nourse–driven regional cooperatives were originally founded to achieve scale economies or provide missing services in contrast to the "additional–margin"–oriented Sapiro regional commodity marketing cooperatives.

7. New Generation Cooperatives. Currently there are only a few new generation cooperatives. They are the result of collective action–oriented founders attempting to address market failure situations, excess supply price depression, cooperative property rights structural weaknesses and free rider issues. Specific solutions in the form of asset appreciation mechanisms, liquidity creating delivery right clearinghouses, proportional patronage distributed control, base equity capital plans and membership policies controlling entrance, are established in their by–laws and operating practices. The initial organizers are as investor–driven as they are user–driven in adapting their financial and governance policies.

146

St. Paul Pioneer Press   April 11, 1994 [Headline]

"KEEPING THINGS SWEET AT CRYSTAL SUGAR"

By Lee Egerstrom
Staff Writer
(Excerpted)

The North American Free Trade Agreement is a done deal. Liberalized trading rules under the General Agreement on Tariffs and Trade are striking down trade barriers and farm subsidies. And Cuban leader Fidel Castro isn't getting any younger.

What this all means is that American Crystal Sugar, the flourishing farmer–owned co–op based in Moorhead, could be in trouble soon unless it makes some big changes.

Political change in Havana could mean an end to the U.S. embargo against importing Cuban sugar. Government sugar policies that have put floor prices under U.S. produced sugar are no longer secure in Washington.

"We've got to be ready for whatever comes," says veteran sugar industry leader Aldrich Bloomquist, who organized sugar beet farmers in the Red River Valley of Minnesota and North Dakota to buy American Crystal more than 20 years ago and turn it into a co–op.

With that necessity in mind, American Crystal is indeed preparing for change. President Joe Famalette and the board are searching for ways to expand membership and sugar production. One possibility they have explored – so far without results – is bringing Canadian farmers from nearby Manitoba into the co–op, either directly or through joint ventures.

Beyond expanding its primary business – which already accounts for 70 percent of the sugar produced in the Upper Midwest – the company also is preparing to diversify. It is in the process of investing in a new co–op that will manufacture high–fructose corn syrup, a rival to sugar, so it can offer a broader line of sweeteners to customers.

*

American Crystal's sense of urgency about changing to keep current with the world food industry comes despite its status as one of Minnesota's most successful food companies.

American Crystal returned $284.1 million in direct payments to its 2,200 grower–members in the 19 Red River Valley counties last year, for an average payment of more than $129,000 per member. North Dakota State University studies show American Crystal operations and the purchases of its farmer–members have about a $1.4 billion annual impact on the valley's economy.

While such bold changes as a move into vegetable processing would come somewhere down the road, the American Crystal board already has approved strategic plans it hopes will propel the co–op into the 21st century.

On Jan. 1 this year (1994), for instance, American Crystal pooled its sugar production with that of Minn–Dak Farmers Cooperative of Wahpeton, N.D. and Southern Minnesota Beet Sugar Cooperative at Renville, Minn., to form United Sugars Corp. The join venture marketing business, headquartered in Edina (Minn.), projects first year sales at $700 million.

That was just a start.

On Feb. 17, Famalette and the American Crystal board chairman, Robert Nyquist, announced board approval of a series of strategic measures. Among them:

* The co–op will explore additional capital improvements, including the possibility of importing new Japanese technology, to increase the extraction of sugar from the molasses byproduct of beef processing. Using current technology, the co–op did expand molasses desugarization, as it is called, at its East Grand Forks, Minn., factory this winter.

The bottom–line purpose of this move is production efficiency, says Famalette. "With all things being equal, including the technology, there is no way a well–run cooperative shouldn't be the low–cost producer."

* American Crystal is working with the Minn–Dak cooperative and state corn growers groups to help start a new co–op, Northern Corn Processors Inc., which will produce high–fructose corn syrup. The new company is now "subscribing," or lining up charter members, in the southern Red River Valley area of Minnesota and the two Dakotas. Farmers pledge corn produced from a fixed number of acres as payment for stock in the company.

An agreement among the new co–op and the two sugar companies calls for American Crystal and Minn–Dak to invest as partners in building a processing factory, called a wet–milling plant, somewhere near the valley.

American Crystal initially will manage the corn fructose plant under contract with the corn co–op, just as it initially operated Southern Minnesota after Renville area farmers started their sugar co–op in the mid–1970s. The fructose sweetener will be marketed through the United Sugars joint venture, expanding its product lines and giving it more clout in the sweetener markets.

"The term we use is leverage," said Robert Atwood, a former executive with the Imperial Holly sugar company in Texas, who returned to his native Twin Cities in January to head United Sugars. "Having fructose means we can go to commercial accounts and take care of all their sweetener needs."

* On another path, the board also has approved an increase in the sugar beet acreage that can be planted this year. And it plans to sell more American Crystal stock to broaden its membership next year.

These two moves are related, and they also tie in with an emerging problem that has resulted from the co–op's success. Stock value in the co–op has climbed faster than the stock of most public companies traded on Wall Street.

When Bloomquist organized the co–op, farmers bought $100 worth of stock for each acre of land on which they wanted to grow sugar beets. The land itself was not part of the deal; rather, each share represents the right to sell the production of sugar beets from one acre. The initial subscription began in 1972; the deal was completed and the company began operating as a co–op in 1973.

Now, in the wake of successes with production and marketing as well as steady improvements in beet seeds and farming practices, American Crystal's stock has been trading at the equivalent of $2,100 an acre, says Mark Dillon, co-op public relations manager. Some reports indicate trades between farmers for as much as $2,300 an acre.

"It's a real problem for new farmers, or not well-off farmers, to pay those prices and become members of the co-op," Dillon acknowledges. The situation creates "political problems" between sugar beet farmers and their neighbors outside the co-ops, adds Bloomquist, now retired from American Crystal. He says the co-op must find ways to keep "doors open to new members."

In a dramatic move last year, Famalette asked economists and farm management specialists at North Dakota State University and the University of Minnesota to help devise ways new farmers could buy shares without deflating the value of existing co-op stock.

He acknowledged the predicament at a meeting late last year with University of Minnesota economists and Twin Cities food company officials. "Let's see hands," he said. "Any of you have problems seeing farmers as millionaires?"

That is a farm problem the University economists don't usually have to deal with. Now they, along with bankers at the St. Paul Bank for Cooperatives and AgriBank, are exploring possible lending programs to make shares of American Crystal affordable to farmers who are just starting out or, if established, are not wealthy enough to buy into the co-op otherwise.

Steve Taff, an agricultural and applied economist who co-hosted the Famalette luncheon, said American Crystal represents one aspect of a problem that is becoming fairly widespread in agriculture. "People just can't buy in and start farming at today's land costs," Taff said. "So we keep coming up with interest subsidies and other schemes to help beginning farmers."

One of the co-ops first steps toward enlisting new members comes this spring. American Crystal will increase production of sugar beets by 20,000 acres, expanding the beet crop to 420,000 acres. An acre of land, which is about the size of a football field, will produce about 19 tons of sugar beets in a typical growing year.

Instead of allowing only members to grow these beets, the co-op will allow some nonmember neighboring farmers to rent a quota for a certain number of acres. And starting later this year or next, American Crystal will develop a system for members and nonmembers to buy these new planting rights as shares in the co-op.

Like a public company, a co-op can split stock or have additional offerings of stock. By expanding the acreage, American Crystal effectively is having a secondary stock offering.

American Crystal's board last year claimed the right to approve or reject sales of stock between farmers, says Famalette. "They don't want the shareholders living in Arizona and Florida and leasing their acreage," he said. "They want this to remain a producer co–op."

To meet that goal will take good planning, say co–op officials. And a certain amount of luck with timing would help.

Bloomquist knows about timing. He was executive director of the Red River Valley Beet Growers Association in the early 1970s, when the Denver–based parent of American Crystal started restricting planting and phasing out processing in the valley.

"We were losing our sugar industry, so we offered to buy the company," he recalls. "They agreed, probably to get us off their backs. I don't think they believed we could do it."

The farmers' takeover coincided with a global drought that reduced sugar production in Asia and Europe at the same time hurricanes swept through the Caribbean, damaging sugar cane fields in Cuba and Central America. U.S. sugar prices jumped from about 13 cents a pound to nearly 70 cents a pound – the highest price in modern history.

"A lot of our early subscribers paid off their investment in the co–op with their first crop," says Bloomquist. "It was incredible timing."

Famalette and the board are well aware of timing, too. They are trying to stay several steps ahead of sweeping changes that are occurring in the world food system.

"(Trade) barriers are coming down. Price supports for all farm crops aren't safe anymore. Farmers can't just produce commodities. They have to move a few steps up the food chain to make sure they share some of the income and profits from processing, not just from production," said Famalette.

He came to American Crystal after serving in executive positions with International Multifoods' prepared foods division in California, and has been in the consumer foods industry for more than 20 years.

"I know the food industry from the brands back down," he said. "Co–ops and farmers have known the food industry from production up. We need expertise from both ends of the food chain in co–op management today to survive the global market."

That means American Crystal must compete in hiring managers with Multifoods, General Mills and Pillsbury, not just with fellow co–ops Land O' Lakes, Cenex and Harvest States, he said.

That managerial experience can direct more diversification and move the farmers and their co–ops up the ladder in the food system, he said. It will make U.S. co–ops more like their European counterparts, which process about 70 percent of food products in northern Europe and more than 90 percent in Denmark and the Scandinavian countries.

"If you look at the brands backwards, you can see what is happening in the food industry and what's happening to farming," Famalette said. "Farmers have to ask themselves if they want to just be farmers or if they want to be farmer–businesspeople.

"If they want to stay a farmer, they will be growing crops under contract for someone else. If they decide to be farmer–businesspersons, they will be growing things for themselves."

A pre–revolutionary barn and a bust of Lenin – neither of which is functional today – are all that remain of the former Zarechenskoe state farm near Moscow. Developing a new, integrated farm–to–market system for private farmers are Valery Shilin, the Zarechenskoe farm manager; Victor Storozhenko, technical advisor from a Moscow agricultural academy; and Ralph Hofstad, former chairman of Land O'Lakes who is mustering American technical assistance through the CUGM organization.

CHAPTER 6:
TECHNOLOGY & KNOWLEDGE TRANSFER

Finding Lessons in What We're Telling Others

> Halsa dem Darhemma,
> Halsa far a mor.
> Halsa grona hagen
> Halsa lillebror
>
> *trans.* – "Greet all those at home,
> Greet father and mother,
> Greet the green meadow,
> Greet little brother."
>
> *trans. (second stanza)*
> "Swallow, think of me also,
> I would like to fly with you.
> The meadow must be green at home,
> Oh, swallow, hear my plea."
>
> – From a folk song of the Scandinavian immigrants. (Swedish lyrics.)

Talk about an information superhighway. All the Scandinavian immigrants had this wonderful song instructing a bird to fly home and deliver greetings to relatives left behind. It is beautiful music that still has strong ethnic appeal in Scandinavian settlements in America, long after the lyrics have lost their emotional meaning for later generations.

Ancher Nelsen, who left Denmark as a young child when his family moved to Minnesota, sang the song beautifully. He assimilated well, or as much as a Dane needs to in Minnesota; becoming a state senator, lieutenant governor, administrator of the Rural Electrification Administration in Washington, and a congressman from southern Minnesota for parts of three decades. I took Congressman Nelsen to dinner at a Capitol Hill restaurant one night after a long evening debate on the 1973 farm bill. He returned the favor at the Republicans' private Capitol Hill Club the following night, when debate went even longer. The club's piano player noticed the congressman's arrival and started playing Halsa dem Darhemma, prompting an unlikely duet, one voice a clear, on–key Danish, the other a muffled Swedish.

Later, at dinner, Congressman Nelsen told a story that explains much about the newness of people in the Northwest states and their bonds to the larger world. President Richard Nixon sent him, Vice President Spiro Agnew and Central Intelligence Agency chief Allen

Dulles as the official U.S. delegation to the funeral of Denmark's King Frederik IX. Mr. Dulles had worked with the Allied underground out of London during World War II and helped Frederik and Danish officials with resistance, intelligence and the smuggling of Danish Jews to Sweden during the German occupation. Somewhere over the Atlantic, the congressman excused himself from the other men and sat alone in the back of Air Force One. The CIA chief and the vice president later found him sobbing in his seat.

"Dulles was scared I was having a heart attack," the congressman recalled. "So I had to tell them. It was just sixty years before that I was down below, crossing the Atlantic in a freighter, going to America. Now I was going back the other way, in Air Force One, to represent my new country at the king's funeral."

The Dulles family arrived in America aboard the Mayflower. Vice President Agnew's father was an immigrant from Greece. Both men understood the congressman's emotions, and what this meant about America, the land of opportunity. They stood in the aisle of Air Force One, hugging each other, and wiping their eyes. "That had to be the most tears on Air Force One since that awful trip from Dallas," said the congressman, referring to the flight to Washington after the assassination of President Kennedy.

Two Michigan congressmen – Gerald Ford and Guy Vander Jagt – came through the Capitol Hill Club dining room that night and saw Congressmen Nelsen and his guest getting misty eyed as the congressman recalled the trip. "Ancher, is everything O.K.?" asked the concerned future vice president and president. "Yes, Gerry, we're just sharing a good memory," was the response. The interruption was enough for teller and listener to regain their composure. The immigrant congressman had one final comment, "You know, I hope it isn't long before the Fourth District (in Minnesota) is represented by one of those Hmong or Mexicans coming into St. Paul." [6–1]

Halsa dem Darhemma.

კ)ෂ෩

Today, Northwest Airlines and its international travel partner, KLM Royal Dutch Airlines, have the "birds" carrying greetings and expertise from the Northwest region of the United States to distant lands. US West, the regional telecommunications company, is another important player in this linkage of the region with all other areas of the world. I see this linkage every day in my work as a journalist.

Former officials from Land O'Lakes, for instance, routinely commute between the Upper Midwest and large state farms in the former Soviet Union that are being converted into production and processing cooperatives. Another former Land O'Lakes executive regularly commutes to Jamaica, where he serves on a board of directors for an economic development fund he helped start. Lawyers expert in cooperative law have been helping new parliaments and farm leaders abroad prepare cooperative laws and business structures to speed modern development. The state Farm Bureaus, Farmers Unions and National Farmers organizations in the Upper Midwest and Northwest states are exchanging farmers with developing countries abroad, and former Farm Credit System bankers from the region are helping establish cooperative lending institutions in former Communist countries. And then there are the academics from private colleges, state university systems and the land grant universities, from Wisconsin on out to Washington and Oregon, who are combining their own research with giving technical assistance and transferring expertise to developing nations, farms and communities throughout the world.

All manner of the seven types of cooperative structures described in the preceding chapter are being developed in the world's new lands. Many are hybrids and cross-bred institutions using several of American cooperative structures. Others have to be called new generation cooperatives because they use every known co-op device to build entire farm-to-consumer food systems to fill existing voids.

Helping hands and expertise are being provided by Americans from all across the land. The Northwest's contribution is noted here because it exceeds demographic proportions, and because the people of the Northwest region are passing along knowledge, technology and business acumen brought to the prairies by our own settlers and refined, over time, in our own communities.

We are generous with the knowledge and technologies we have developed or refined on the American frontier. A lot of it is imported from Europe and "homelands" of our immigrants. We've adapted it. Some of it is our own invention, built from seeds our immigrants planted. No matter the origin, we are now sharing both expertise and technologies with a new generation of pioneers and community builders around the world who are trying to bring their resource-based communities into the modern world.

Americans should stop and listen to what their own people are saying and study what we are sharing with others abroad. We, too, have towns that are dying and serving no more important economic function than yesterday's state farms in Russia. We may find ways to revive rural America by heeding our own words. Examples of history-changing development projects, many involving people from the Northwest region, will be offered later in this chapter. But first, attention should be given to why people from the Northwest reach out to others in developing countries.

જાજાજા

The Northwest has always been connected to the outside world, through the mails as immigrants corresponded with relatives, through railroads and steamships that supported trade, through today's airplanes and fiber optics. These connections make an information highway, no matter how rustic it was or sophisticated it's become. It is an especially well–traveled highway right now.

Within a six month period in 1993, for example, I talked with people who had recently worked on economic development projects or research projects in Russia, the Ukraine, Kazakhstan, Uzbekistan, Pakistan, Hungary, the Czech Republic, Slovakia, Poland, Estonia, Latvia, Lithuania, Morocco, Thailand, the Philippines, Malaysia, China, Columbia, Peru, Ecuador, and the occupied territories on the West Bank and Gaza. Separately, I visited or corresponded with researchers back from Australia, New Zealand, Costa Rica, Jamaica, Sweden, Germany, Nigeria and Brazil who were studying ways to improve farming practices and agrarian economic developments.

This global involvement doesn't easily fit with the Northwest's image as the spiritual home of America's isolationist political beliefs. These beliefs were especially prevalent before World War I and World War II, as political scientists and political historians have written time and again. There were, however, economic and sociological aspects to that isolationism that have been largely overlooked in big, national studies, but more carefully recalled by regional historians. Unless the regional economic and sociological aspects are understood, it is impossible to understand the region's past or its people today. Simply put, the Northwest never held what economists might call a "fortress frontier" mentality in which its people believed they could live and prosper while disconnected from the larger world.

This is shown in the 1972 book, <u>Almost to the Presidency,</u> in which author Albert Eisele recalls the time two future U.S. senators, Hubert H. Humphrey of Minnesota and Russell Long of Louisiana became a college debate team that stomped a visiting team from Oxford University. Humphrey and Long were graduate students at Louisiana State University. The debate topic, on the eve of World War II, was, "Should the U.S. go to the aid of Great Britain in the war in Europe?" [6–2] Considering what we know today about those two LSU graduate students, one has to feel pity for poor little Oxford. Its team didn't have a chance. Later, when asked about that debate, Humphrey recalled that

156

it was fun. "I think that was the only time I ever gave an isolationist speech." [6–3]

The former vice president and long–time senator was masterful with words. His graphic depiction could risk angering people, at times, if they looked too narrowly at what he was saying. On one such occasion he was describing the importance of trade on the Minnesota economy. "Without it, we would be nothing more than 'East Dakota,'" he said one night at a party with Ridder Publications reporters in Washington. Then remembering we reporters also corresponded for newspapers at Aberdeen, South Dakota, and Grand Forks, North Dakota, he said, "Let me put that another way..."

No offense was intended. What the native of Doland, South Dakota, meant was that his adopted state of Minnesota and the entire surrounding region are totally dependent on trade linkages and economic growth in other countries and regions of the world.

This dependency started from the earliest days of settlement. It continues today, because the Northwest economy needs the larger world market as much as ever. But there is more involved; the American agrarian spirit of settlers and early builders lives on. People closely tied to the land and its resources see new frontiers emerging and are drawn to them, to lend a hand with development planning and technology transfer. The land, it seems, nurtures people who become active in various cooperative, government and voluntary agency development programs abroad.

Yes, the Northwest was home for much of the isolationist political sentiment that marked the first half of the Twentieth Century when the world was going to war. And yes, most academic studies of the isolationist movement focus on political, social, and sometimes anti–Semitic attitudes of the people. These political and social studies aren't wrong. They are, however, incomplete. Historians such as Robert P. and Wynona H. Wilkins from the University of North Dakota, for instance, have shown the linkage between local economics and political thought, putting the Northwest's moments of isolationism in more understandable light.

The Wilkins, in their bicentennial history, North Dakota, explain nearly all Northern Plains politics, including the creation of North Dakota's NonPartisan League, in the context of a wheat economy. Wars are hell on trade, they noted. Immigrants left their homes in Europe to settle the Northern frontier, in part, to avoid those almost endless European wars. Moreover, the militaristic German governments before the two world wars were diverting the German domestic economy to the then–arms race. Germany was actually a good customer for Northern Plains grains. [6–4]

That doesn't justify the isolationism that swept through Northern state parties and agrarian politics in the first half of the now-concluding century. But it does fill historical voids left by the volumes of studies and nonfiction books that simply look at isolationism as a political movement or social philosophy. It should also be noted that once the decision to go to war was made, North Dakotans led the nation in voluntary enlistment, as a percentage of population. [6–5]

The memory of isolationism also ignores the strength of the region's political leadership of the past full century. Montana's Mike Mansfield, the intellectual Senator leader and later ambassador to Japan, was not an isolationist or a parochial wheat and cattle lawmaker from Big Sky Country. Nor was Washington state's Senator Henry "Scoop" Jackson. Senators Hubert Humphrey of Minnesota and George McGovern of South Dakota, often working in league with Senate Foreign Relations Chairman Frank Church of Idaho, could overcome some constituent doubts and lead efforts to pass foreign aid bills. Senators Warren Magnuson of Washington and Milton Young of North Dakota, the chairman and ranking Republican of the Senate Appropriations Committee, respectively, made sure there was adequate funding for foreign aid development and food assistance programs during the post–World War II era. [6–6] Members of the House of Representatives from the Northwest states, while not as prominent except for House Speaker Thomas Foley of Washington, were equally bold in arguing for foreign aid programs and appropriations. Indeed, Republican Congressman Mark Andrews of North Dakota moved to the Senate and carried on the appropriations legacy of Senator Young after the senior senator's death.

There was one person who bridged both eras from isolationism to active internationalism. Senator Quentin Burdick, son of a Republican and NonPartisan Leaguer, Congressman Usher Burdick, was a Democratic–NonPartisan League senator from North Dakota for nearly 30 years. On several occasions during the 1970s, the senator told me how he visited the battlefields of India and Pakistan after border battles in the 1960s. There were American–made tanks and equipment burned up and lying all around. The United States had armed both sides. "I decided then that I would never vote for another military aid bill, and I never have." [6–7] He did, however, support foreign economic assistance and food aid bills when Congress and administrations didn't combine those appropriations bills with military aid.

Senator George McGovern of South Dakota also recalls the friction between India and Pakistan, having shipped surplus food aid and powdered surplus milk to "West Pakistan" while he was Food for Peace director for President Kennedy. Later, in 1971 when West

Pakistan became an independent Bangladesh, the United States again shipped surplus dry milk to the war–torn country, though many people in Bangladesh had a lactose intolerance and couldn't consume our surplus without getting sick. Others did get sick; most likely from mixing the powdered milk with contaminated water from municipal wells and water systems damaged and neglected during the war. "I've often wondered if the real reason Bangladesh didn't go Communist was because there was no one healthy enough to lead an insurrection. It is the only country that spent its first six months as a nation suffering from severe diarrhea. Of course, they called it 'Uncle Sam's revenge.'" [6–8]

As a child, I heard Senator Humphrey proclaim, "Hunger, starvation anywhere in the world, represents a lost market for the American farmers." [6–9] It was true then; it is true now. It was a rallying cry for farmers to support foreign assistance measures that would, in turn, build markets for themselves. It was an effective message, if now recognized as a somewhat shallow message; but Senator Humphrey wasn't a shallow man. Economists and other social scientists who work on development issues refer to that message as "enlightened self–interest." There are compelling moral issues at stake, as well. Moreover, foreign aid too closely linked to building markets for farm trade have often lost sight of development objectives. [6–10] Senator Humphrey knew how to use enlightened self–interest arguments to muster support behind public policy issues that weren't popular in many parts of the country.

This reaching out to the greater world didn't start with the generation of leaders represented by the Mansfields, Jacksons, Churches, Humphreys, McGoverns, Magnusons and Youngs. It was the environment years before by frontier business leaders who knew they had to reach out for markets; they lacked convenient local populations to consume their goods and commodities. [As discussed in Chapters 1, 3 and 7] It was not essentially different than the historic, commerce–driven view of the world shared by industrialists and shippers along America's coasts. Senator McGovern, in his 1977 autobiography, Grassroots, tells of presidential candidate John F. Kennedy struggling with an agricultural speech at the National Plowing Contest at Sioux Falls, South Dakota, in 1960. But later that same day, speaking extemporaneously at the Corn Palace in Mitchell, South Dakota, Kennedy was in full flower. "I don't regard the ... agricultural surplus as a problem. I regard it as an opportunity ... not only for our people, but for people all around the world." [6–11] What would play in Boston harbor during Ambassador Joseph Kennedy's trading days played well for his son at the Corn Palace.

This was the environment for farmers, townspeople and industrialists in the Northwest to grew up, work and prosper. Political leadership contributed to it. But it doesn't explain it. American democracy doesn't allow political leadership to get too far in front of the governed. Civil rights took 100 years from the Civil War before they started winning both constitutional and statutory protections under Presidents Eisenhower, Kennedy and especially Lyndon Johnson. The health care reform initiatives promoted by President Clinton come nearly 40 years after President Eisenhower toyed with similar ideas. Farm programs, it should be obvious to most agrarian observers, almost always deal with world market conditions of the past, rarely anticipating the future. A constant throughout, however, is the dependency the Northwest region has on the outside, bigger world – the new play on University of Wisconsin historian Frederick Jackson Turner's "safety valve" theories on the frontier that were discussed in earlier chapters. It is a constant because it serves the region even when public policies fail. Let us look, then, at some of the new pioneers who are recycling our regional expertise and technologies out to the world's new frontiers.

ଔଔଔ

Ralph Hofstad, former president and chief executive officer at Land O'Lakes, commutes regularly between his consulting office in Edina, Minnesota, and temporary offices near Moscow. From both points, the successful agribusiness leader is helping Russian farmers, educators and community leaders plant the seeds for a modern food and agribusiness system.

Larry Buegler and Burgee Amdahl, the two immediate past presidents of the Farm Credit Bank of St. Paul, now known as AgriBank FCB, have also been regular commuters between Russia and Poland and their homes in Minnesota and Wisconsin. They worked out of offices in Moscow and Warsaw in the early 1990s where they helped Russians and Poles establish cooperative lending institutions to finance agricultural and rural development. [6–12] At one point in early 1993, Mr. Buegler took 17 bankers and finance experts from Northwest and Oklahoma cooperatives, public and private banking institutions and colleges on two–week and month–long consulting trips to work with Russian and Kazakhstani bankers and farm leaders. [6–13]

A constant exchange of people between the former Soviet bloc countries and the Northwest states began in the late 1980s, with economic reforms under former Soviet President Mikhail Gorbachev, and intensified with the fall of Communism. Just as Nikita Kruschchev insisted on visiting Iowa during his historical visit to the United States, President Gorbachev chose to visit the Twin Cities of Minnesota three decades later to meet with agribusiness officials from the Midwest. Both visits went to the heart of Soviet and Russian needs, though both seemed to puzzle writers in major financial centers of the country. [6–14] These needs haven't changed. When the winter snows were melting in early 1993, state Farm Bureau organizations were lining up host farms for visiting Russian and former Soviet nationals to learn American farming techniques with a season of on–the–job training. Parliamentarians, government planners and farmers visited campuses, attended training sessions and led fact–finding missions from Madison, Wisconsin, on out to Corvallis, Oregon and Pullman, Washington.

When the Northwest's linkage to the former Soviet and Eastern bloc states weren't direct, the region's farmers and related experts worked through national organizations. The National Grain and Feed Association – one of America's oldest trade associations though it is often confused with the Feed Grains Council – contracted with the U.S. State Department's Agency for International Development (AID) and started a farmer–to–farmer exchange in early 1993. It, in turn, contracted with the National Farmers Union and National Farmers Organization to recruit American farmers for the three–year project. And with the help of additional private foundation grants, the project would keep a continuous supply of farm and agribusiness men and women serving as volunteer technical advisors at development projects in the Russian Federation, Kazakhstan, Kyrgyzstan, Tajikistan, Turkmenistan, the Ukraine and Uzbekistan. [6–15]

Winrock International Institute for Agricultural Development, an extraordinary research foundation started by a branch of the Rockefeller family at Morrilton, Arkansas, and the Center for Agricultural and Rural Development at Iowa State University, were subcontractors on the U.S. AID project. "The (grain and feed organization) is extremely pleased to be asked to be a conduit through which these republics of the Commonwealth of Independent States can gain access to a wide range of agribusiness expertise available within our industry," explained NGFA President Kendell W. Keith in announcing the program. While the cooperating farm organizations mustered farmers to work with farmers in the CIS states, Mr. Keith's group was finding expertise among its 1,200 grain, feed and processing companies to work with CIS cooperatives and new business ventures.

"The NGFA and its members have tremendous credentials to offer in assisting the former Soviet Union (to improve) its grain storage, handling and processing system; in establishing cash and futures markets; in setting up a system of trading rules; in using arbitration as a means of resolving trade disputes, and marketing and transportation of agricultural commodities. All sectors of U.S. agriculture –indeed our entire nation – have a great stake in assisting the CIS in developing a market–based agricultural system and a sounder economy under democratic institutions." [6–16]

Events from 1989 on in the former Soviet Union and neighboring countries are extraordinary and were unimaginable a decade earlier. Events at home, in the United States, were also extraordinary, as shown by agriculture's response to farm and food needs in the former Communist countries. Though not all Americans share the views expressed by Mr. Keith, agriculture appears to be reaching a consensus that poverty abroad is harmful to farmers' economic health. This wasn't so widely understood outside the Northwest region of the country as recently as 1987 when some agricultural trade associations were still fighting foreign aid economic development appropriations in Congress. [6–17] In the 1990s, the debate inside agriculture is more properly focused on what shape foreign development aid should take. [6–18] No one seems to argue the old notion that we can use hunger and starvation to lock customers into our markets. [6–19]

ಐ ಐ ಐ

Regardless of future U.S. public policies and public opinion, proximity to the land does encourage people to do what is right. Economic self interests, foreign policy objectives and moral imperatives blend together. There is a response to need. "You can't throw the former state farms up in the air and expect them to come down as cooperative or private farms," explained Mr. Hofstad after returning from one of his Russian trips in the summer of 1993. Many of the early Russian attempts to "privatize" the former state–owned, or collective farms, were destined to failure, he says. Some farms were declared cooperatives. The new farms got the equipment left over from the state enterprise, but then started losing state subsidies that were needed to keep the previous farms afloat. Moreover, simply declaring a state farm a cooperative didn't do anything to make the farm a viable business unit. Often, he said, these farms were expected to support a community of workers and their families that could not be sustained by the farms' resources.

It didn't take an experienced American agribusiness executive long to see that economic democracy in the Russian Federation would need a complete overhaul of the food and agribusiness system, from the farms on through to the retail marketplace. It took Mr. Hofstad one inspection visit.

162

The Russian development plan began in a simple fashion. Churches United in Global Mission, an interdenominational assistance program founded by the Rev. Dr. Robert Schuller of the Crystal Cathedral in Southern California, sent food aid shipments to Russia in 1991 and early 1992. After that experience, members of the Russian Parliament asked Dr. Schuller in June, 1992, for American help in developing private farms. Dr. Schuller is a native of Iowa and knew there were better places in America to look for farm expertise than in Orange County, California. He called the Rev. David T. Scoates, head pastor of the large Hennepin Avenue United Methodist Church in downtown Minneapolis.

Pastor Scoates thus became chairman of the CUGM Russian Farm Project Committee. He enlisted Mr. Hofstad as executive director, and the co-op leader, in turn, enlisted help from Vern Moore, a former colleague and senior executive at Land O' Lakes. They then enlisted the help of Dr. Lee Kolmer, the retired dean of agriculture at Iowa State University. [6–20] In short order, American agricultural experts were doing an assessment of Russian needs and resources available for the project. Pastor Scoates explained why. "It's as if a great uncle who owned all the land died without leaving a will." [6–21]

While Pastors Schuller and Scoates made organizational arrangements with members of the Russian Parliament, the Russian Orthodox Church and Russian farm organizations such as AKKOR – the Russian Association of Private Farmers and Cooperatives – Misters Hofstad, Kolmer and Moore went searching for academic and technical partners. The found another Iowa educator, Dr. J.T. Scott, who was a visiting Fulbright professor at the Timiriazev Agricultural Academy in Moscow. He opened doors to Russian academics from Timiriazev who could be brought into the project.

By late 1992 and early 1993, a strategy was in place for CUGM's Russian Farm Project using American farm and agribusiness models, North American and Western European extension service models, French cooperative models and new, international agribusiness models that have no national origin.

At the ground level, CUGM operatives and their Russian partners planted demonstration plots on May 1, 1993, using land provided by a newly formed joint stock company farm and the Russian Orthodox Church. [Chapter Note A] The Timiriazev Agricultural Academy would later disperse information from the plots to other farmers through an extension service patterned after the American model. And on another level, the Americans were making plans to help from 20 to 25 farm families "homestead" new private farms on about 4,000 acres of state lands that were made available to the project. Each farm family would start with from 120 to 150 acres of land while sharing labor and equipment. Expertise gained in this homesteading project would also be shared through the new Timiriazev extension service apparatus.

The size of these farms would leave the new Russian homesteaders about 50 years behind their American, Canadian, Australian and Brazilian counterparts, assuming they will gain access to modern farm equipment and technology. So while the initial seeds were being planted in 1993, Mr. Hofstad and his collaborators were at work on developing the next stage of the food system, just as progressive communities in rural America are now doing. "There were a lot of woods around the potato farm where we were working," Mr. Hofstad recalls. "I asked our Russian friends to start looking for small sawmill opportunities. We can use the resources available to build better storage on the farm and in the distribution system. This could reduce storage losses 40 percent to 50 percent."

These steps, while absolutely essential for Russian agricultural development, are merely a beginning. They will still leave Russia's new generation of private and cooperative farmers decades if not a century behind their agrarian colleagues in developed countries. To move up the food chain and provide the farmers with value–added income, the Russian Farm Project has launched feasibility studies for building a butter and milk powder processing plant that would use milk from private and cooperative farms. And Mr. Hofstad is using his international contacts with cooperatives to explore possible joint venture business deals that would aid technology transfer and help generate Russian farm income. Among the first people he sought out for exploratory talks was I.J. Prins, chairman of the large Cebeco Handelsraad cooperative in the Netherlands. Under study is a potential joint venture potato processing enterprise with Cebeco.

Neither processing project was a definite deal at the time of this writing. But Mr. Hofstad says such enterprises must be started to assure private farmers a market for their production and a share of the marketing profits. What's more, he has an international model – and one that has been slow to arrive in the United States – for making Russian processing enterprises work.

 හශශහ

The Russian project follows a model that comes from the southwest of France, where farmers belonging to Coop de Pau took some bold risks beginning a sweet corn business in 1976. Pau is about midway between Toulouse and the Cote Basque, near Lourdes. It is a land that has produced good corn crops since the early 1960s. The farmers

164

decided to expand into sweet corn to diversify their cooperative while developing an alternative crop for their farms. What they needed was the marketing experience to bring sweet corn into the mix of popular vegetables in the French and European markets.

The farmers and managers of Coop de Pau went looking for business partners who had the marketing expertise they lacked. They found the Pillsbury Company in Minneapolis. Pillsbury was the parent of Green Giant, a vegetable subsidiary that is the leader in canned and frozen sweet corn marketing. Green Giant and Pillsbury have since been taken over by London–based Grand Metropolitan Plc., which has added to the French farmers' marketing strength on their continent. But before Grand Met arrived on the scene, the farmers of Coop de Pau and Pillsbury had formed a joint venture processing company, Seretram, that became the largest sweet corn canning business in Europe. [6–22]

Following this model, Coop de Pau has since created a second vegetable joint venture processing business with France's Bonduelle company. Called SOL (Sud Ouest Legumes), this joint venture processes green beans and broccoli as well as sweet corn at two plants in Borderes and Labenne. And on different agribusiness paths, Coop de Pau has expanded into seeds and plant breeding in a joint venture with Monsanto and in a vegetable oil processing joint venture with The Andersons. [6–23]

"I've been watching Coop de Pau grow for the past 10 years," Mr. Hofstad said during a mid–1993 interview. And, like Mr. Prins in the Netherlands (See Chapter 4), the U.S. cooperative official kept his ties with new–found friends in France to learn from their experiences in creating a new generation of co–op businesses. This linkage started one year when Coop de Pau asked its joint venture partner, Pillsbury, for help in finding an annual meeting speaker. Pillsbury executives in downtown Minneapolis called across town to Land O'Lakes, in the St. Paul suburb of Arden Hills, offering Mr Hofstad a trip to the south of France. "They twisted my arm," he jokingly recalls. The then-chairman of Land O'Lakes went off to Pau where he told farmers about trends and developments with North American cooperatives. That started a 10–year friendship in which the French have continually shown him new ways for cooperatives to do business, he said.

What has come from that relationship is a joint venture model that Mr. Hofstad is sharing with the Russians in the CUGM project, and with U.S. and Canadian cooperatives that seek his expertise. A brief sketch of the model follows:

Production – Growers or their cooperatives should control the land resources, contracting production, supplying farm seeds, chemicals and fertilizers, and the employees needed in production.

Processing – The growers should have control over employees and support staff for the joint venture processing company, with the marketing company partner having responsibility for management of the joint venture. Both the growers or cooperative and the marketing company should have an equal equity investment in the joint venture. Though not all business officials would agree with this, Mr. Hofstad is adamant about shared, equal ownership. "From what I've seen in business, owning 49 percent of something means you don't own very much," he says. "When ownership is 50–50, both partners are committed to making the joint venture work."

Marketing – The marketing company partner owns this segment of the shared enterprise and provides management for marketing. Ideally, however, Mr. Hofstad says the growers should have a future option to buy a minority stake in the marketing company. While that might be an ideal, it is not part of Coop de Pau's arrangement with its partner. The farmers of southwest France are not likely to buy an equity stake in the publicly–traded Grand Met, a diversified food and beverage company with annual sales of $12.5 billion. [6–24]

ഇ‍ഇ‍ഇ

Since radical change began in the former Communist, or centrally planned countries in the late 1980s, economists have rediscovered a phenomenon that has affected most countries in periods of rapid economic change. Reform doesn't instantly make standards of living rise and national economies prosperous; rather, reforms and change usually lead to a period of economic decline and a lowering of standards of living during periods of adjustment.

This creates a challenge for international institutions and volunteers from abroad. Patience will be rewarded, but there will be few immediate successes. Technology and knowledge can be quickly transferred from the First World to the Third World or new, developing former Communist countries. This "things often get worse before they get better" phenomenon has held back both political and economic reforms in Russia and some other nations of the former Soviet Union. This delay threatens political stability in some countries, as noted in a particularly important paper, **"Economic Reforms in Poland: Implications for Agriculture,"** published in 1993.

Written by Harald von Witzke and Ulrich Hausner of the University of Minnesota's Center for International Food and Agriculture Policy and Janusz Chichon of the Olsztyn University of Agriculture and Technology, Olsztyn, Poland, the paper notes the dual challenge facing all the former Communist countries: decentralizing their economies and institutions at the same time they seek reintegration into the European and world economies. "Poland has chosen a strategy of reform that is now commonly referred to as economic shock therapy. The country went through a major decline in economic performance but it is now recovering at remarkable rates," they wrote. In Poland, "Unlike many of its neighboring countries, attempts at economic and political stabilization have, by and large, been successful. This is an important precondition for keeping Polish capital in the country, for attracting private investors from abroad and for obtaining assistance from the international donor community. Much remains to be done, however, to assure that Poland remains on track toward sustainable economic prosperity." [6–25]

Richard Magnuson, a career cooperative business law expert currently with the Midwest law firm of Doherty Rumble and Butler, has watched political delays hold back even the fast–changing Polish economy. He worked with Polish agricultural leaders and members of Poland's parliament to propose a model business code for converting state farm property into Western–style cooperatives. Though Poland and Hungary are regarded as the fastest adapters of economic reform among former Communist nations, Mr. Magnuson watched, from afar, as his cooperative business model was amended and adopted piecemeal. At first, the delay was frustrating, he recalls. Then a certain realism set in. Nothing happens quickly in Washington, D.C., either. In fact, the Warsaw government moved more quickly, as it turns out, than Washington's usual pace for handling economic and social programs. [6–26] One potentially controversial item in his proposed model was making long–time employees of state agribusiness companies and state farms stakeholders in the buyout of the property by cooperative operators and farm investors. In time, the Poles agreed and fashioned a program in which employees would be paid for their sweat equity in new cooperatives. While Polish government leaders, economists and lawyers studied his proposals, Mr. Magnuson served as a similar volunteer consultant for the new government of independent Estonia. Copies of his model business law proposals have also been forwarded to Latvia and Albania. [6–27]

Mr. Magnuson, a former senior vice president and general counsel at Land O'Lakes, was among more than 250 farmers, cooperative business executives and academics sent to former Communist countries as technical advisers and consultants between 1989 and 1993 through the offices of the Volunteers in Overseas Cooperative Assistance of Washington, D.C. VOCA, as the organization is known, is a non–profit development assistance program supported by major U.S. agricultural, consumer and finance cooperatives, with financial support from the U.S. State Department's Agency for International Development. Founded in 1970, VOCA has worked in more than 70 countries around the world. Gary Slaats, manager of VOCA's Upper Midwest office at Madison, Wisconsin, notes that more than 1,000 volunteers were scheduled to work abroad during 1993. Their expenses are paid; their other rewards are emotional. While Mr. Magnuson was helping newly independent countries of Eastern Europe, for instance, three other volunteer attorneys from Iowa, Washington state and Virginia worked with Albania to prepare articles of incorporation for cooperatives. Ten other cooperative law specialists have advised the Russian government on drafting co–op business laws.

Much of the work is of a technical nature, such as legal work to accommodate the conversion of state property into private farms and cooperatives. Much of the hands–on, technology transfer information, however, is geared to helping farmers and peasants pool resources to operate modern hog farms, poultry farms, dairies, vegetable and fruit processing and marketing cooperatives, and information exchange systems. In short, these programs are using American expertise to find ways to avoid the long evolutionary process of agrarian development that has occurred from frontier days to the present in the United States and Canada.

But that is only one program in which American cooperatives are helping transform newly independent countries, developing Third World countries and newly industrialized (NIC) countries that seek help with agrarian development. Most of America's major regional cooperatives are members of Agricultural Cooperative Development International, another Washington D.C.–based economic development and cooperative education nonprofit organization. It was formed in 1968 by the merger of two predecessor groups, the International Cooperative Development Association that had been formed by co–ops with closer ties to state American Farm Bureau Federation organizations, and the Farmers Union International Assistance Corporation that had been formed by the National Farmers Union and its cooperative creations.

Today, most of the nation's largest farm, credit and insurance cooperatives are active members of ACDI. [See Chapter Note B] It works with the Agency for International Development and cosponsors on development projects around the world. It also operates development programs supported by Public Law 480, the so–called Food for Peace program. ACDI, in turn, is also a sponsor of VOCA. [6–28] Bill Black, ACDI assistant vice president for communications, said recent development projects range from establishing an agribusiness information program for Egyptian television, "Serr El Ard," or "Secrets of the Land," to helping create a rural farm credit system in Mexico. [6–29]

ဆဆဆ

Cooperative leaders, so capable of lending a hand to others abroad who are looking for ways to improve their standards of living, access to credit and markets, and agrarian communities, are well aware of similar needs at home. They have taken steps to make their own cooperative businesses more efficient. They are creative in forging partnerships and linkages with each other and other companies when cost–effective and practical. They are heeding their own words.

These steps are not enough. But they are a promising start.

A number of "hybrid" agribusiness companies have been formed in the United States since the mid to late 1980s in which public, private and cooperative companies have crossed lines of incorporation to form joint venture enterprises. Archer Daniels Midland has numerous joint venture businesses, as noted in Chapter 2. Harvest States Cooperatives has all three types of incorporated business partners in the A.C. Toepfer grain trading joint venture company at Hamburg, Germany, and is a partner with a French, family–owned company in its malting barley processing business. Land O' Lakes spun off its fluid milk and juice business in the Upper Midwest market into a publicly traded company that it continued to control, though it has since bought that company back and integrated it into the cooperative. And more recently, Land O' Lakes placed its animal health products operations in a joint venture with private investors. That joint venture intends to become a biotechnology company with both animal and human food and health products, and is expected to make an initial public offering of its stock in the mid–1990s.

More such business arrangements are inevitable. It allows groups of growers and cooperatives to gain access to marketing and other business expertise they don't have at hand. Now, with North American and international trade agreements removing trade barriers and protections for American producers, there is a new urgency for being creative. "It's a global economy," says Mr. Hofstad. "You have to produce for the market, and you have to help make that market."

There is an undeniable urgency for change to bring the Russian Federation and developing countries out of one era and into another, bypassing a long, evolutionary peasant culture by creating integrated farm–to–consumer food systems.

I would argue that a similar urgency exists in rural America, the Prairie Provinces of Canada, rural Europe and everywhere else that has

169

a highly developed but government–dependent agriculture system. Granted, it's a tough sell trying to convince successful farmers they must make changes and expand their stake in the food system's chain, or ladder. This is especially true after two or more decades in which agricultural production and financial success have been closely tied to crop specialization and monoculture agriculture. But models for change are available to guide American farmers, agribusiness leaders and rural community planners. We have helped create them.

Halsa grona hagen. Little birds can cross oceans, flying in any direction.

Chapter 6 FOOTNOTES

1. These personal recollections and interview comments were intended for publication at the time of Congressman Ancher Nelsen's death in 1993. However, since the worst corporate communications are usually found within communications companies, this material never got used.

2. Eisele, Albert. Almost to the Presidency: A Biography of Two American Politicians. The Piper Company. 1972. Page 50.

3. From an unforgettable luncheon conversation in the Senate Dining Room at the U.S. Capitol Building. Betty South, the senator's press secretary, reminded the former vice president that April 10, 1975, was the Egerstrom's first wedding anniversary. My wife, Lalinda LaMotte, was press secretary for Sen. Henry Bellmon, Sen. of Oklahoma at the time we were married. Senators Humphrey and Bellmon, Vice President Nelson Rockefeller and House Rules Committee Chairman Ray Madden of Indiana had cleared logistical hurdles for us to get married. Humphrey promised to take us to lunch on our first anniversary; he did.

4. Wilkins, Robert P. and Wynona H. North Dakota. W.W. Norton & Co. 1977. For a good discussion of isolationism, see Chapter 9, **On Guard Against the "Warlords."**

5. Ibid.

6. The author reported on their work with foreign aid and appropriations bills in Congress during the 1970s. He observed that senators and members of Congress from other regions would support these bills, but they hesitated to take leadership positions on foreign aid matters when their constituents had problems at home.

7. From news accounts, Grand Forks Herald, Grand Forks, N.D.

8. From news accounts, Aberdeen American News, Aberdeen, S.D., December, 1974.

9. This is from memory; exact wording cannot be found. It was a radio broadcast of a Humphrey speech delivered at an annual meeting of Farmers Union Grain Terminal Association, now known as Harvest States Cooperatives.

10. For a comprehensive look at "enlightened self–interest" and its role in U.S. foreign assistance, see: Ruttan, Vernon W. Why Food Aid? The Johns Hopkins University Press. 1993, and Third World: Customers or Competitors? E.A. Jaenke & Associates, Washington, D.C. 1987.

11. McGovern, George. Grassroots. Random House. New York, N.Y. 1977. Page 82.

12. Former bank presidents Buegler and Amdahl were not alone in working on agrarian financial reform in the former Communist states. A June 23, 1993 statement issued in Washington by Agricultural Cooperative Development International announced that Don Theuninck, vice president for Audit and Review Department, AgriBank FCB, was honored by the organization for his volunteer work in helping reform the farm credit system in former Soviet Union states.

13. From news accounts, St. Paul Pioneer Press. 1993.

14. There is nothing like a visit from a prominent head of state to reveal American ignorance of economics and geography. Any visit between New York or Washington and California, with the possible exception of a stop in Chicago, triggers all manner of essays in journals seeking answers to "Why?" It may not have happened yet, but the day will certainly come when a headline will ask if the king, president or prime minister is lost.

15. National Grain and Feed Association announcement, February 16, 1993.

16. Ibid.

17. Third World: Customers or Competitors? E.A. Jaenke & Associates. Washington, D.C. 1987.

18. Ruttan, Vernon. Why Food Aid? Johns Hopkins University Press, Baltimore, Maryland. 1993.

19. Third World: Customers or Competitors? For examples, see Chapter One observations by Ward Sinclair, the Washington Post: Lee Egerstrom, St. Paul Pioneer Press; and Loren Soth, the Des Moines Register; and Chapter One development arguments by John Block, former Secretary of Agriculture; James Houck, University of Minnesota economist; and Robert Paarlberg, Wellesley College political scientist.

20. For reports on CUGM Russian Farm Project plans, see: Dawson, Jim. "Russian farm project to have local support." (Minneapolis) Star Tribune. January 15, 1993; and Morphew, Clark. "Church group will help Russians learn how to farm independently." St. Paul Pioneer Press. January 15, 1993.

21. From organizational announcement, Churches United for Global Mission. January 14, 1993.

22. Coop de Pau annual report. 1990.

23. Ibid.

24. Hofstad, Ralph. From papers prepared for Coop de Pau and Russian farmers.

25. Von Witzke, Harald; Chichon, Janusz; and Hausner, Ulrich. **Economic Reforms in Poland: Implications for Agriculture.** (Monograph) Center for International Food and Agriculture Policy, University of Minnesota. 1993.

26. Magnuson, Richard. From papers prepared for the Polish government and for the Volunteers in Overseas Cooperative Assistance, Washington, D.C., and from conversations over lunch between 1990 and 1994.

27. For an outstanding portrayal of modern, post–Communist Poland, see Pichaske, David. Poland in Transition. 1989 – 1991. Ellis Press. Granite Falls, Minnesota. 1994.

28. From annual reports. Agricultural Cooperative Development International. Washington, D.C.

29. As reported in "ACDI Signs First Protocol with Mexico," Cooperative News International, a publication of Agricultural Cooperative Development International, Washington, D.C. Volume 6, No. 4, 1993.

The agreements creating the Zarechensky Joint Stock Company farm, which is working with the Churches United for Global Mission project in Russia, is described in the following paper provided by V. Storozhenko, agricultural advisor and professor at Timiriazev Agricultural Academy, Moscw:

Zarechensky joint stock company was established by general meeting of shareholders 12.15.1992 on the base of former state farm. It was registered in accordance with resolution of the Dmitrov district (Moscow region) Administration Head, signed on 12.31.1992.

The company consists of 264 shareholders. Every member of the community has his own capital in the form 4.13 hectares of land and amount of money from 40 to 120 thousand rubles in prices of 1992. This amount of money depends on the term of their employment in the former state farm.

The joint stock company main specializations are production of milk, potato and feed, including forage grain.

The company has 1,500 hectares of cultivated lands, 600 dairy cows, 600 head of young stock, 57 tractors and trucks and other agricultural machinery. It also has service center and woodworking shop.

Joint stock company works in tight contact with local farmers, helps them with tillage, supplies them with seed, fertilizers and pesticides, (and) helps to sell farmers production.

Amount of village inhabitants – 2,500, including 300 children and 200 retirees.

Joint stock company has eight departments: crop production (35 employees), animal husbandry (73), garage (44), workshop (18), building department (13), kindergarten (14), dwelling service (17) and administration (28).

The company is located in 25 km from Dmitrov, 20 km from Dubna, 35 km from Klin, 40 km from Sergiev Posad, 40 km from Istra and 95 km from Moscow.

Zarechensky joint stock company has no special buildings for marketing.

Chapter 6 CHAPTER NOTE 6–B

Member organizations of the Agricultural Cooperative Development International, as listed in the development group's 1991 annual report, included:

Farm Supply, Processing and Marketing Cooperatives
Agricultural Council of California California. Trade association. Sacramento, California.
Agway Inc. Farm supply and food marketing co–op. DeWitt, New York.

Associated Milk Producers Inc. Milk marketing and processing co–op. San Antonio, Texas.

Blue Diamond Growers. Almond marketing and processing co–op. Sacramento, California.

Cenex. Farm supply and petroleum co–op. St. Paul, Minnesota.

CF Industries. Fertilizer manufacturing and marketing co–op. Long Grove, Illinois.

Countrymark Cooperative Inc. Grain marketing and farm supply co–op. Indianapolis, Indiana.

Dairylea Cooperative Inc. Milk marketing cooperative. Syracuse, New York.

Farmland Industries. Farm supply, petroleum and food processing co–op. Kansas City, Missouri.

Growmark Inc. Farm supply, grain marketing and processing co–op. Bloomington, Illinois.

Harvest States Cooperatives. Grain marketing and processing co–op. St. Paul, Minnesota.

Indian Farmers Fertiliser Cooperative Ltd. Fertilizer manufacturing and farm supply co–op. New Delhi, India.

International Cooperative Petroleum Association. Suppliers of lubricants to member co–ops, and provider of petroleum services. White Plains, New York; Dordrecht, the Netherlands; and Paris, France.

Land O' Lakes Inc. Dairy, food processing, marketing and farm supply co–op. St. Paul, Minnesota.

Maine Potato Growers Association. Marketing and supply co–op. Presque Isle, Maine.

MFA Oil Company. Petroleum supply co–op. Columbia, Missouri.

Mississippi Chemical Corporation. Fertilizer manufacturing co–op. Yazoo City, Mississippi.

National Cooperative Refinery Association. Petroleum producing and refining co–op. McPherson, Kansas.

Norpac Foods Inc. Fruits and vegetables processing and marketing co–op. Slayton, Oregon.

Seald–Sweet Growers Inc. Citrus marketing co–op. Vero Beach, Florida.

SF Services Inc. Farm supply co–op. North Little Rock, Arkansas.

Southern States Cooperative Inc. Manufacturing and marketing co–op for farm supplies, fertilizers and petroleum. Richmond, Virginia.

Sunkist Growers Inc. Citrus processor and marketer. Van Nuys, California.

Tennessee Farmers Cooperative. Farm supply co–op. LaVergne, Tennessee.

Tri Valley Growers. A fruits and vegetables processing and marketing co–op. San Francisco, California.

Universal Cooperatives Inc. An interregional manufacturing, importing and buying co–op. Bloomington, Minnesota.

Farm Credit Banks
CoBank – National Bank for Cooperatives. Denver, Colorado.
Farm Credit Bank of Baltimore. Baltimore, Maryland.

Farm Credit Bank of Louisville. Louisville, Kentucky. (Now merged into AgriBank FCB; St. Paul, Minnesota.)

Farm Credit Bank of Spokane. Spokane, Washington.

Farm Credit Bank of Springfield, Springfield, Massachusetts.

Farm Credit Bank of St. Louis. St. Louis, Missouri. (Now merged into AgriBank FCS, St. Paul, Minnesota.)

Farm Credit Bank of Texas. Austin, Texas.

Farm Credit Bank of Omaha. Omaha, Nebraska. St. Paul Bank for Cooperatives. St. Paul, Minnesota.

Western Farm Credit Bank. Sacramento, California.

National Organizations

National Cooperative Business Association. Washington, D.C.

National Council of Farmer Cooperatives. Washington, D.C.

National Farmers Union. Denver, Colorado.

National Grange. Washington, D.C.

Insurance Companies

MSI Insurance. St. Paul, Minnesota.

Nationwide Insurance. Columbus, Ohio.

The small shops, nearby agriculture, and value–added processing such as wine making, brewing, meat packing, baking and light manufacturing, make the Amana Colonies in Iowa a leading Midwest tourist attraction and a model for community collaboration.

Photo courtesy of the Amana Colonies Convention and Tourist Bureau

CHAPTER 7 THE MORE THINGS CHANGE
Lessons From Alternative American Communities

Thanksgiving

We believe in passion:

    as flatlanders
    we know all things
    great and small come
    from passion;

We cherish the holy magic of touch:

    as animals
    we are all joined
    by this electricity;

We believe in the poetry of wave collisions:

    as people from
    diverse countries
    building a national unity

We are grateful as well:

    to all sacred
    powerful
    omnipotent

    gods who thankfully, this year
    left us be.

        Donal Heffernan. Orion.
        Lone Oak Press. 1993.

There are parallels at home to the new Russian and Eastern European communities that are now seeking ways to adjust to changing times. They have one major thing in common. A sense of community that doesn't have to be established or revived.

Some of America's most successful rural communities were founded by people who brought pre–existing bonds to their new settlements. Like the Pilgrims before them, these people were a community long before the first log house was built and before the first furrow was plowed. And other community builders were the people already living on the frontier, the Native Americans, who built villages to serve their tribal needs and support commerce. In some cases, whole communities picked up and moved to new lands to escape religious and political persecution or to find land for expansion.

In most of these cases, including the giant westward movement of the Mormons into Utah, an original intent was to remove the religious followers of a group from the pluralistic society that was evolving in most parts of the United States. With a few notable exceptions, such as the Old Order Amish, most of these groups changed over time. Radically.

Followers of the Church of Jesus Christ of Latter Day Saints, or Mormons, now send missionaries throughout the world on proselytizing missions. And Mormon communities in Utah send trade missions across North America and throughout the world seeking trade partners and outside investors to keep building the Utah economy. Success on either front leads the Mormons into more integration with the broader, pluralistic American society and the world in general.

Native Americans, meanwhile, were forced into resettlements while others sought refuge on reservations from the encroachment of settlers. In some cases, the new communities were forced on the tribes after they were removed from valuable resources, such as the Black Hills. Regardless of the origins of reservation communities, few tribes have sought to be totally isolated from the larger environment. [See references to the Navajo in Chapter 2, Chief Wabasha and the French in Chapter 3.] In the 1980s and '90s, many tribes were again reaching beyond their tribal boundaries for new wealth and capital, building a tourism economy centered around casinos and Indian gaming businesses.

Great changes are occurring, yes. But the people who seek to build strong communities around ethnic or religious affiliations are successfully making changes to preserve and build from their earlier foundations. Rural communities can learn much from these groups as community leaders seek guidance for coping with change and for positioning their communities and agrarian countryside for the next century. These alternative, or atypical style communities, should be studied because their methods for surviving and thriving in a constantly changing world may well have application for more heterogeneous communities. They change, but they stay a focused community.

మమమ

The Amana Colonies

For tourists to Middle America, one of the great memories is a walk down mainstreet of Amana, or "Main Amana," as it is known in Iowa.

The street is lined with quaint brick and stone houses and commercial buildings that date back to the Civil War era. Inside, the buildings serve as bread and breakfast lodging houses, antique and craft stores, old world German and American restaurants serving family–style, country fare, and various retail stores offering gift merchandise, specialty meats and food products including the colonies' wines and beer.

All seven villages that make up the Amana Colonies of east–central Iowa deserve a recreational visit. Main Amana is cited here only because it has more of the charming businesses and old buildings that make the colonies Iowa's most popular tourist attraction. [7–1] To visit the colonies right takes two to three days; the traveler with only one free day should park and walk the streets of Main Amana.

Though started as an agrarian commune with town crafts people to support a self–sufficient, isolated lifestyle for the colonies' religious followers, the Amana villages are today an extremely successful tourism venture of cooperating merchants, farmers, artisans, professional people and manufacturers. Rozella Hahn, spokeswoman for the Amana Colonies Convention and Visitors Bureau, said Iowa sales tax receipts show tourism generates about $30 million annually for the seven villages' enterprises. These sales generate $1.25 million in sales tax revenues for the state, which is equal to half of Iowa's total tourism promotion budget. Ms Hahn also concludes that tourism sales also provide a payroll of $17 million for the 1,620 people employed in the villages to serve tourists. [7–2]

How the Amanas were founded, and how they have changed, offer lessons for rural community planners who now know their communities must change to assure their continued existence in the next century.

The Amanas trace their history to southwestern Germany where Eberhard Ludwig Gruber and Johann Friedrich Rock formed a charismatic splinter group from the Lutheran Church in 1714. They found converts to their new church, the Community of True Inspiration, in Switzerland and France as well. Following the path of other American immigrants, later leaders took their followers to Buffalo, New York, in 1842 to escape religious and political persecution in Europe. They bought 5,000 acres of land in the Seneca Indian Reservation near Buffalo to start their Ebenezer community.

Land needs for their people quickly had Inspirationist church leaders exploring for a new home on the frontiers of Kansas and Iowa. They chose the beautiful Iowa River Valley that is now the Amana Colonies. But before leaving New York, the Inspirationist people started a rare but successful adventure with what we would later call Communism. [7–3]

Many of the immigrants who came to join the Ebenezer community in New York were impoverished and couldn't afford to buy their own farms and shops. Church leader Christian Metz proposed a constitution for the community that adopted a communal system of ownership. By accepting this system of communal living, church followers determined that living within the community was of greater value than acquiring personal property and personal wealth.

Beginning in 1855, groups of people from the Ebenezer community began migrating to the new villages built in the Iowa River Valley. They gave their villages and settlement the name Amana from the biblical Song of Solomon, since Amana means to remain faithful, or true. About half of the 1,700 residents of the Amana Colonies remain true to the founding church, which is now more conveniently called the Amana Church, says Mike Shoup, secretary of the Amana Society and director of its agribusiness operations. The other residents are Lutherans, Catholics, Methodists and other religious followers typically found in the pluralism of eastern Iowa.

A bigger change, however, came to the economic and social structure of the colonies. Three years into the Great Depression, and with more social and commercial interaction with neighboring Iowa communities, the people of the Amanas voted in 1932 to abandon their communal way of life. This decision, which gave shape to the modern Amana Colonies, is still called "the Great Change" by people in the region. Capitalism, in nearly all its forms, was reborn.

At first glance, the transformation of the Amanas would seem to be a perfect model for adapting state farms and enterprises in the former Soviet bloc countries to free enterprise capitalism. At the same time, the economic vitality displayed in the Amanas should serve as a model for small town community leaders throughout North America.

Most communal property was sold to families living in the Amanas. Communal kitchens became bakeries. Shopkeepers bought their shops. Professional people, though they had been sent away to colleges and universities by the people of the Amanas, became private practice professionals like their counterparts in other Iowa towns. Certain large revenue producing ventures were kept by the community with ownership transferred to a newly formed private business corporation called the Amana Society Inc.

The Society continues to have about 700 shareholders, says Mr. Shoup. Outsiders buy into it "whenever they find someone willing to sell stock," he said, though it doesn't happen with the regularity of most publicly traded companies.

What remains is a stock company that is primarily community—owned and operated to keep businesses operating and profitable within the community. In many respects, it is similar to the way in which farmers pool resources to start agricultural processing businesses that are usually structured as cooperatives. Amana Society holdings are broader than most farm cooperatives, however, and include the Amana Furniture Shop, Amana Meat Markets, a Holiday Inn, pharmacies and general stores, the Amana Woolen Mill, and 27,000 acres of adjacent farmland.

It is only the latter enterprise – farming – that would be politically unacceptable in most communities of the Middle West and Northwest. Private ownership of farms is too honored an American tradition to follow the Amana example. Mr. Shoup has about 25 full–time farmers who take care of the small grains, soybeans, corn, vegetables and forage crops while tending a 23,000–head beef herd and a 220–sow hog operation that farmers call "farrow to finish" because they breed the sows and feed the pigs until they are finished, or ready for market.

With that exception noted, the colonies' Amana Society does offer a useful example of how rural communities can band together with town and country capital to provide plants, employment and amenities of life that they would not likely have with traditional entrepreneurship and private capital. The Amanas are making use of a traditional form of investing – a stock corporation – for nontraditional purposes – community enterprises and development. At the same time, they are making use of a more traditional source of capital that is also serving the community well.

This latter form of capitalism is outside ownership, or "foreign ownership" as long–time Amana families might view it. It involves one of the two original woolen mill sites. Located in the village of Middle Amana, the woolen mill was closed in 1937 and converted to an electrical company. It was renamed Amana Refrigeration Co. in 1950, by which time the name Amana was a nationally recognized brand name for refrigerators and household appliances. The company was sold in 1965 to the Raytheon Corp., which had the national presence and wherewithal to compete with the General Electrics and rival manufacturing giants.

Today, this Amana subsidiary of Raytheon is the biggest employer in the colonies, says the Society's Mr. Shoup. The factory in Middle

Amana provides employment for between 2,500 and 3,000 people from the villages and surrounding towns.

This mixture of ventures and community solidarity gives the Amana Colonies a vibrant economy that mixes tourism with agriculture, recreation with manufacturing. All 475 buildings from the original village building projects are listed on the National Register of Historic Landmarks. "We have about 20 businesses that are still owned by the Society," says Ms Hahn from the convention and visitors bureau. "And we have 85 businesses that were bought, or pulled away from the Society's businesses." With some of the best restaurants and river valley scenery to be found in Middle America, these latter day Amana capitalists have achieved the near utopia their church leaders set out to create in the New World.

ಬಂಬಂಬಂ

Mennonites

These derivations of the Sixteenth Century Anabaptists deserve a fresh look by community planners throughout the Northwest states. Their followers today are more than good neighbors for people living in rural areas of the Northwest; they are experienced community builders.

Mennonite communities can be found reaching out from Nebraska into various areas of the Middle West, Southern Plains and Prairie Provinces of Canada. While strict adherents favor plain clothes and simple living, they are not usually as conservative in their lifestyles as the Amish, for whom they are often mistaken. Actually, the Amish are a more isolated sect that broke away from the Mennonites in the Seventeenth Century. Their church leaders, or elders, keep in close communications with the Mennonites in the United States.

Despite plain living, the Mennonites make good use of modern agricultural technology. And they reach out to the world. A Mennonite farmer from Nebraska was consistently on loan, or in his case a mission, during the 1970s to work with the Inter–Religious Task Force in Washington, D.C., and with similar ecumenical groups in recent years. This is recalled because the Mennonite farmers, whose farms were cared for by neighbors during their absence, were invaluable sources of expertise for government officials involved with foreign assistance programs. Though not interested in holding public office themselves, the Mennonite farmers were willing citizens who would step forward and give advice. This was important during most of the 1970s and 1980s when farm organizations were willing to ship food aid assistance abroad but were less inclined to support technical assistance programs. It fell to the religious community, working together, to keep American foreign aid programs focused both on feeding the hungry world and helping less fortunate people find ways to grow food. [7–4]

I admired the work of the Mennonites on world hunger and development issues in Washington, D.C., during the 1970s, as noted in Chapter 3. Yes, they have built communities to live and work with one another somewhat removed from the world around them. But they never ignored the greater world and its problems. They are, as we are told in various ways by our religions, their "brothers' keepers." I was continually impressed with their social responsibility actions during the late 1980s and early 1990s, when I served on the board for the Commission for Church in Society of the Evangelical Lutheran Church in America. And in January, 1994, I saw their brilliance and decency being put to work on human and community needs in their own backyard.

Mountain Lake, Minnesota, is primarily a Mennonite community. It is located near larger towns that have meatpacking, agricultural processing and manufacturing jobs. The changing farm economy and the farm depression of the mid–1980s thinned out Mountain Lake's business community, and it caused a huge outmigration of residents, leaving boarded up homes and a depressed local housing market. Laotian immigrants discovered the inexpensive housing and the nearby jobs; the American dream was reborn and Mountain Lake became home for about 200 new residents.

Calvin Beale, chief demographer for the U.S. Department of Agriculture in Washington, D.C., said the immigration of people to Mountain Lake is part of a new wave of immigration discovering rural America in various parts of the Middle West, South and Southeast seaboard states. [7–5] The key, of course, is the availability of rural job opportunities – there no longer is inexpensive, virgin land available for settlement on the frontier. Mountain Lake, with strong civic and school leadership and financial and moral leadership from its seven churches – five of which are Mennonite congregations – has to serve as the model for rural communities that are going through both social and economic change. [See Chapter Note A]

ജ്ഞജ്ഞ

Hutterites

While the Mennonites originated in the Netherlands, the Hutterites were founded in Moravia. Now, 400 years later, most Hutterites in America are of German origin and continue to use the German language nearly as often as English within their communities. Most live communally in communities stretching across the Dakotas, Montana and in the Alberta and Saskatchewan provinces of Canada. "It's a simple lifestyle, but not like the Amish or some of the more conservative Mennonites," says T.J. Gilles, the long–time agriculture writer for the Great Falls Tribune in Montana. "If there is a new hog confinement barn pictured in a magazine, the Hutterite carpenters will be busy drawing it to scale and you'll see a new barn going up in a few days."

Their handiwork is legendary from the Dakotas to the Canadian Rockies. Their community building, however, is genius.

To care for their people, the Hutterites can't simply expand their land holdings and increase plantings of wheat and barley crops as is the pattern for most grain farmers in the Northern Plains. Rather, they realize, livestock raising is the first step in value–added processing and community–building through employment. So the Hutterites maintain a dairy industry in Montana that is far from dairy country, they herd beef cattle and geese, they have poultry and egg operations even though they live far from major urban centers, and they have large hog farrowing and feeding operations that are the match of those found in Iowa and the Southeast seaboard states.

"You can almost say the Hutterites own Montana's dairy industry," says Mr. Gilles. [7–6] There are a few other dairy farmers around, he says, so they have teamed with the Hutterites to own and operate a dairy processing and marketing cooperative.

Livestock, meats and eggs are a bigger marketing problem for these Montana producers, Mr. Gilles said. Those products are shipped into California and Pacific West Coast states and are increasingly finding export outlets to Asia and Mexico. Shipping livestock products out of expansive Montana isn't easy, compared to bulk grain shipments. But the Hutterites have taken the extra steps necessary to do it. "All the Hutterites can't be grain farmers," says Mr. Gilles. "Even Montana isn't big enough for that."

No place is anymore. So the Hutterites are now building industrial communities for their followers, recognizing the physical and economic limits to agriculture for their growing membership.

A Hutterite community near Brookings, South Dakota, for instance, has helped establish such an industrial community at Dexter, Minnesota, that is now known as the Oakwood Bruderhof. Clark

Morphew, the highly regarded religion writer and syndicated religion columnist at the St. Paul Pioneer Press, noted in a 1992 feature on the community that the community rents its 700 acres of prime farmland to neighboring farmers. [7–7] Meanwhile, the community's workers prepare, plane, glue and finish maple boards and panels that are sent to another Hutterite community in Pennsylvania. That community uses the processed wood products to build toys, furniture and equipment for day–care centers and schools that are sold commercially under the brand name "Community Playthings." [7–8]

ಐಐಐ

The Amish

At first glance, the Amish communities of Ohio, Pennsylvania, Wisconsin, Iowa and Minnesota would seem to have little to offer "the English," or non–Amish people, living around them and in neighboring communities. That would be an understandable but erroneous impression. The Amish need to be extremely creative to keep their old ways and survive economically in the modern world.

They are, without question, the pioneers of what people call "sustainable agriculture" and "low–input agriculture." They don't use electricity, tractors or trucks. This creates problems for the Amish from their base in Lancaster, Pennsylvania, on out to newer settlements in Ohio and the central states. It is a special challenge, however, for the Amish farmers in the "Bluff Country" regions of southwestern Wisconsin, northwestern Iowa and southeastern Minnesota. If God intended this rolling countryside along the Mississippi River for any purpose besides hardwood forests and tourism, it would have to be pastoral agriculture. Cows can graze the lush hillsides; farmers must be careful how they plow and till the soil, if they do it at all, to avoid soil erosion. Farm chemicals must be carefully used or they seep through the karst rock formation quickly into farm and town water supplies.

Given that geographical environment, dairy farming and dairy processing are big business in this three–state region of the Upper Midwest. But to be a Grade A milk producer, a dairy farmer must have electricity and refrigerated bulk storage tanks. The Amish use old fashioned milk cans that sit in well water for cooling overnight. That means their milk is classified as Grade B, suitable for manufacturing into cheese, ice cream or other products, unsuitable for the fluid, or drinking milk market.

By 1990, Minnesota had only two rural creameries, or milk gathering points, that still handled the old fashion milk cans. Both were in southeastern communities and Amish farmers were their main suppliers.

"The future wasn't looking good," recalls Minnesota Agriculture Commissioner Elton Redalen, who farms near the Amish at the Bluff Country community of Lanesboro. "Nobody wants to handle milk cans anymore when bulk tanks and trucks are so efficient." It was questionable how much longer their local creamery would keep accepting cans.

The Amish contacted their cousins and brethren in another Old Order Amish settlement at Cashton, a small community near the river city of La Crosse, Wisconsin. Those farmers formed Hill & Valley Cheese Co–op in 1983 when they faced similar uncertainties over outlets for their milk. The Wisconsin Amish thus assured themselves a milk market by starting a cheese plant that would handle milk cans.

The Wisconsin Amish lent their Minnesota cousins a hand. They put the Minnesotans in contact with Bob Kiesau, owner of the Kickapoo Business Services Co. at La Crosse. His company provides business services, such as accounting, for Amish communities in several states. "The Minnesota Amish weren't as liberal as the Cashton farmers and they didn't want to form a co–op," Mr. Kiesau said in an interview. [7–9] They were willing to make a financial commitment to assure themselves of a market that would keep them in the dairy business.

Working with Amish farmers, the Minnesota Department of Agriculture and Lanesboro city officials, Mr. Kiesau formed a partnership with Mike Everhart, the cheesemaker at Hill & Valley, and with Nancy Martinson, a La Crosse television station executive. They bought an abandoned manufacturing plant in Lanesboro and remodeled it into a cheese factory. The Amish farmers supplied carpenters and laborers for the task. They also provided the partnership with a business loan of about $75,000 for startup costs. The loan was equal to about $1,000 for each farm that would supply milk to the cheese plant.

"We needed the commitment that we would get the milk supply," said Mr. Kiesau. So the loan agreement put conditions on the Amish as well as the cheese factory investors. It guaranteed the supply of Amish milk by stipulating that $1,000 of the loan would be forgiven for each Amish farm that pulled out of the supply arrangement or quit the dairy industry.

Susan Bain, bookkeeper for the Fremont Co–op Creamery at nearby Fremont, Minnesota, said her cooperative sold the Lanesboro partnership its milk can handling equipment and milk cans in September, 1991. With the start of the Lanesboro cheese operation, Fremont was able to abandon the cans and handle milk exclusively from bulk tank trucks. "But I have to say I do miss seeing the Amish come into town with horses and wagons to deliver some of their cans," she said.

With the transfer of the equipment, River Valley Cheese Co. at Lanesboro was born. Randy Haakenson, the cheesemaker, carefully watches cheese form from milk and starter cultures in three open vats, using modern stainless steel equipment but ages–old cheesemaking processes. He works in front of big–walled windows that allow tourists and curious shoppers to watch the cheese take shape. Lanesboro, a tourist town along the Root River tributary of the Mississippi, bears a resemblance to an Amana Colonies village. Now it also has a gourmet cheese factory and shop for tourists near its local winery and community theater.

It's a perfect setting for the beautiful irony that River Valley Cheese has come to represent. It is a maker of specialty and old fashioned cheeses gathered in cans from hand–milked cows on Amish farms. At the same time, it is succeeding in the marketplace by being a modern, value–added business carving out a profitable share of a niche market for specialty cheeses. "It really is a beautiful combination of the old and the new," says Agriculture Commissioner Redalen.

That still leaves about 50 Amish farms along the southern Minnesota and northern Iowa border whose milk cans are delivered to the area's last can handling creamery. Susan Phillips, office manager of Granger Farmers Co–op Creamery at Granger, Minnesota, said Amish farmers are the primary owner–members of her milk cooperative. Therefore, she said, the cooperative is committed to finding a market for its members' Grade B milk. The Granger co–op now empties the cans into tanker trucks and "English" drivers take the milk in bulk to a nearby Associated Milk Producers Inc. manufacturing plant in Iowa.

Back in Lanesboro, the partnership between private investors and cooperating Amish farmers appears to be working. In 1992, its first year of operation, River Valley sent most of its cheese production to food companies in bulk for repackaging and labeling, said Cheesemaker Haakenson. River Valley's goal is to graduate from the bulk cheese business into premium priced, branded products. It has started an advertising campaign for the River Valley brand. Its first sales campaign was in Amish and Mennonite newspapers that publish in communities across the nation, and distribution of mail order catalogs to Root River Valley tourists who passed through town.

About $40,000 was generated by mail order sales in River Valley Cheese's first year, out of total revenues of about $4 million. That's not a bad start, says economist Dr. Donald Ault at Ag–Nomics Inc.

consulting service in New Brighton, Minnesota. The plant handled about 130,000 pounds of milk a day during its first year, or about half the plant's capacity. In contrast, most modern cheese factories handle about one million pounds of milk a day, the consultant said.

But that smaller size plant can be a strength, not a market weakness, Dr. Ault added. It is about the right size for making specialty cheeses, like those made in Europe, that aren't easily mass–produced in the larger, more automated plants. And it offers a vivid example of how a community of people can turn a market disadvantage into a marketing strength that holds great promise for the future.

ဆဆဆ

Native American Tribes

Similar quick studies could be offered of Indian reservations that combine successful tourism programs and manufacturing with agriculture or mineral extracting. At other reservation sites, case studies can soon be made.

Altin Paulson, vice chairman of the American Indian Chamber of Commerce and a board member of other national economic development organizations involved with Native American and reservation communities, said model, or showcase tribal communities already exist. Alaskan tribes have developed their fishing resources, for instance, and some smaller tribes in Wisconsin have successfully become home to satellite manufacturing ventures that extend out from Milwaukee and Chicago. In some instances, he said, the tribes have invested in joint venture manufacturing businesses with industrial corporations, like ventures described in earlier chapters. Tribes throughout the West are having varied degrees of success in developing and coping with tourism, and developing further, or value–added processing plants, to raise the value of their resources. [7–10]

The Winnibago tribes of Wisconsin, for instance, have several manufacturing ventures on their reservations. They also have a Twin Cities sales office down the hall from Mr. Paulson's business services office in St. Paul. This allows Winnibago sales representatives quick access to Minneapolis–St. Paul International Airport, Mr. Paulson said, and convenient meeting space for sales meetings with business people passing through the Twin Cities and western Wisconsin area.

A big challenge for tribes in coming years, Mr. Paulson says, will be in diversifying local economies that are now gaining a burst of capital from casino and gaming operations. For some, strategically located tribes that are easily accessible to large population areas or served by efficient transportation, gaming income may be enough. But other tribes, more distant from a steady source of customers, may well be hurt in future years as Indian gaming spreads across the countryside.

The more distant tribes will need other sources of tribal income to maintain current standards of living or improve their current status. Moreover, state hospitality industries were already lobbying state governments at the time of this writing, seeking changes in state laws to allow resorts, bars and entertainment centers to cut in on the Indian casinos' action. History does teach painful lessons: We know what happened to the Oglala Sioux when gold was discovered in the Black Hills of South Dakota. Their land –and gold – was taken away.

Mr. Paulson is optimistic that tribes will use gaming income to diversify and build their local economies. For one thing, the Native American business community is better organized now than at any time in past history. And there is far more public and private assistance available for tribal economic development planning. "It's the smell of money," says Mr. Paulson, who operates a successful business consulting service that organizes trade shows, training programs and seminars for corporate America. "Gaming income is making everyone aware of the tribes," he said. "We've become the Native Americans, not the Invisible Americans. That's progress, but we have to keep it going."

ဢဢဢ

Southern Black Cooperatives

There is another group of Americans who must be mentioned, and they have turned to cooperative institutions to combat structural problems with their markets, financing and changing economics. They are the black farm families who work within the Federation of Southern Cooperatives.

The Rev. Will Campbell, a Baptist minister, recalled some of the Federation's history of struggles in the Deep South states in an article for the September/October issue of World Magazine, The Journal of the Unitarian Universalist Association. [7–11] The report notes that Unitarians – Universalists are among socially conscience people who have supported the Federation over the years.

The Federation was created as part of the Civil Rights movement of the 1960s. Its successes aren't as easily measured as other regional cooperatives', if one uses the typical measures of gross sales and membership. But the Reverend Campbell portrays a far more important measurement – there are still black farmers owning their own land in Georgia, Alabama and Arkansas. [7–12]

Looking back, the struggles of black farmers in the South bear kinship with obstacles faced and overcome by pioneers on America's Western frontier. The Federation and its leaders were hounded by people wanting them to fail, whether for old–fashion racial bigotry reasons or by people wanting the small farmers' land. In the Northwest, Indian land owners have endured similar struggles. And the pioneers who formed early cooperatives had to overcome class struggles that attempted to keep farmers in their place..

It won't be dealt with in this book, but it should be noted that founders of what is now Land O' Lakes were jailed in Minneapolis because they were perceived as a threat to the region's economic power trust. Forerunner organizations of Harvest States Cooperatives fought similar battles, in and out of grain markets and in and out of courts. This is stated only for the following point: The Federation of Southern Cooperatives could have a far more important future for its members and their rural communities than its past might suggest. Small farmers in the South, as elsewhere in America, will need to secure more of their family income from value–added processing earnings. And while building a structure for processing and marketing their production, they will also be creating jobs for other community residents and strengthening their local economies.

ഇഇഇ

There is a key element found in all these different communities of Americans that will be discussed in more detail in the next chapter. It is the strong, well–educated local leadership now found in most vibrant communities. These leaders can give guidance to planning and decision–making, and recognize when outside expertise is needed.

I have seen wise planning when the Norwegians sent teams to Kuwait to study how best to use its North Sea oil revenues so it wouldn't destroy the Norwegian work ethic and national culture (See Chapter 3). I've seen large corporations, including Knight–Ridder Newspapers, turn to the Indian business leader, Altin Paulson, for outside consulting services and help with planning a jobs fair. I've seen the Amana Colonies, the Amish and other groups wisely make changes, so they can remain strong, healthy and essentially unchanged. It happens in tribal communities of the West, too, and in black communities of the South. Leadership makes the difference.

ഇഇഇ

A postscript on the linkage between alternative communities and the cooperative spirit:

On May 30, 1994, former Agriculture Secretary Ezra Taft Benson died at Salt Lake City, Utah. In recent years he was head of the Church of Jesus Christ of Latter–Day Saints.

The Mormon leader served two terms as secretary of Agriculture under President Dwight D. Eisenhower in the 1950s. He previously served as executive secretary of the National Council of Farmer Cooperatives from 1939 to 1944, and helped start the Idaho Council of Cooperatives while farming in Idaho.

Chapter 7 FOOTNOTES

1. Rasdal, Dave. Cedar Rapids Gazette. Cedar Rapids, Iowa. 2. Economic briefing report. Amana Colonies Convention and Visitors Bureau, Amana, Iowa.

3. Willkommen. Summer, 1993 edition. A tabloid guide to the Amanas published for the Amana Convention and Visitors Bureau and Amana Society, Amana, Iowa.

4. This observation is based on personal experience working as a journalist covering foreign aid authorization and appropriations bills in the U.S. Congress during the 1970s. Also, former Agriculture Secretary Bob Bergland discussed at length the contributions made by Mennonites, Catholics and mainline Protestant denominations on foreign aid and hunger issues in conversations at the World Food Council meetings in Ottawa, Canada, in 1978. Others contributing to the discussions were Kathy Patterson, then a reporter for the Kansas City Star; former Undersecretary of Agriculture Dale Hathaway; the Rev. Larry Minear, then director of the Inter–Religious Task Force in Washington, D.C.; and Thomas R. Sand, then a special assistant to the Agriculture secretary.

5. Beale, Calvin. Chief demographer, U.S. Department of Agriculture, Washington, D.C. Quoted in "A New Lease on Life." St. Paul Pioneer Press, St. Paul, Minnesota, Jan. 9, 1994.

6. Gilles, T. J. From interview, July, 1993.

7. Morphew. Clark. "Communal Life Rewarding for Hutterites." St. Paul Pioneer Press. Nov. 15, 1992.

8. Ibid.

9. A similar report on the Amish farmers and River Valley Cheese Co. at Lanesboro, Minnesota, appeared in June, 1993, St. Paul Pioneer Press, St. Paul, Minnesota.

10. Paulson, Altin. From interview, December, 1993.

11. Cambill, Will D. "The Federation of Southern Cooperatives: Staying the Course." World Magazine. The Journal of the Unitarian Universalist Association. September/October, 1994.

12. Ibid.

Chapter 7 CHAPTER NOTE 7–A

Reprinted from St. Paul Pioneer Press, Jan. 9, 1994: [Headline]

A NEW LEASE ON LIFE

MOUNTAIN LAKE, Minn. – When the farm depression of the '80s finally ended, Mountain Lake civic leaders wondered if their community would ever recover.

More than 300 residents had packed up and moved away, dropping the population to 1,906 in the 1990 census. Thirteen commercial buildings were boarded up, and 60 homes were vacant.

Three– and four–bedroom homes were available for as little as $3,500, but there were not takers, recalls City Clerk Elizabeth Schmidt. "We were really hurting."

No one foresaw the turnaround that would hit Mountain Lake in the next few years – or the new residents who would spur it.

Mountain Lake's new lease on life came from 54 Laotian immigrant families. Its houses are filled with the new homeowners now, and its schools with their children. The newcomers have opened several businesses in town, and more are in the planning stages.

It's happened quickly, say city officials. Three or four Laotian families moved here in 1989. They told friends and relatives about the city and about job opportunities in the area, triggering a steady migration that now numbers about 200 people.

And they've been joined by a smaller number of Hispanic families who are new to Minnesota, although some were born in this country. All the newcomers have pushed Mountain Lake's population above 2,000 again, said city officials.

Mountain Lake's renewal mirrors what is happening in many other rural communities throughout the Midwest and South, where once–dying local economies based on agriculture are being rejuvenated by Asian, African or Hispanic immigrants.

The newcomers usually are attracted to an area by semi–skilled jobs in such agriculture–related operations as meat packing and food processing. Large numbers already have settled in regional cities such as Worthington, Marshall, Albert Lea, Austin and Willmar (all in Minnesota), where most of those jobs are centered.

But more recent arrivals are gravitating to smaller towns, where housing is cheaper and the rural lifestyle is similar to the one they grew up with. All over the country, such immigrant groups are helping to build a new rural economy from the shell of the old one as they recycle rural homes and mainstreet businesses.

"The people were friendly. Homes were affordable. We like the schools. There was opportunity," explains Siha Douangvilay, an immigrant from Laos who has started three businesses since he arrived in Mountain Lake in 1991 and is considering a fourth.

In other words, says Denny Wilde, city administrator and economic development director, Siha and the other Laotian families were looking for a chance to better themselves in Mountain Lake, just as German and Scandinavian settlers did a century ago.

While Northern European settlers originally came there after cheap farmland to start a new life, the new settlers are attracted by jobs and affordable housing, adds Wilde.

"It's the same American dream," said George Brophy, director of the Development Corp. of Austin, who is seeing a similar migration into southeastern Minnesota.

This marks the first major wave of immigration to rural Minnesota since shortly before World War I, says Tom Gillaspy, state demographer. Much of it has occurred since the 1990 census; this spring, Gillaspy's staff will begin estimating the number of people who have been involved in the moves.

"I wish we could start over and do the census now," Gillaspy said. "This is radical change occurring in rural Minnesota. It is like the great waves of the 1880s, 1890s and 1910 period that settled the Minnesota frontier."

Its new residents won't restore Mountain Lake to the farm service center it was before the farm depression. It has become a bedroom community for those who work in nearby towns.

But along with the laborers for area industries have come some entrepreneurs, too. Siha opened a furniture store in what was a boarded–up former drugstore. He has since added a clothing store that sells both American and imported Asian clothes, and he also operates SH.D. USA–Laos Enterprises, an exporting business that ships late–model used cars from Minnesota to Laos, Thailand and Japan.

Siha said he worked as a car and furniture salesman in Minneapolis and Bloomington after moving to Minnesota from Baltimore in 1982. "Too big," he said in describing Baltimore and the Twin Cities (metropolitan areas). "Everything was expensive. I don't like big cities. Mountain Lake was about right."

City officials are helping Siha find a building with adjacent lot to open a used car business in Mountain Lake. If it works out, the town's existing Chevrolet dealership could also benefit from customer traffic generated by the used car business.

Three other families have opened Asian grocery stores in what had been vacant buildings, although only one of the stores is a full–time enterprise. In another building that once was empty, Chanpheng and Saemgchanh Chantharak have opened the Sunshine Restaurant, which offers Chinese and Thai dinners.

During the day, Chanpheng works as a mechanic at the large Cenex service station in Mountain Lake; his wife began commuting last week to a new job at the Toro lawnmower factory in Windom.

"We were in the restaurant business in Lowell (Massachusetts) before we came here," said Chanpheng.

While there are no Hispanic–owned stores yet in Mountain Lake, the supermarket now carries an assortment of Mexican and other Hispanic food items.

More startup businesses are expected this year. The Rev. Tong Chitchalerntham, pastor of the Lao Christian Fellowship Church, said another family is negotiating to open a bakery business that intends to distribute both Asian and traditional American food products statewide.

The seven existing churches in Mountain Lake pooled resources to bring the pastor and his family here to form the Lao church, which shares facilities with Bethel Mennonite Church. He also serves as an interpreter for new arrivals who need help with English in school and business affairs.

The new migration to rural America is closely tied to changes in the meat packing industry and to the dispersion of semi–skilled manufacturing jobs into rural areas, says Calvin Beale, chief demographer for the U.S. Department of Agriculture in Washington.

"We've seen this around Dodge City, Liberal and Garden City in Kansas, and at a few places in Nebraska and Iowa. Columbus Junction, Iowa, is becoming Hispanic.

"There are places with new broiler industry plants in northern Alabama that now have large Hispanic populations, and around seafood plants in North Carolina and Maryland. And now Minnesota – it really is a trend," Beale said.

At Mountain Lake, for instance, the Laotian and Hispanic newcomers commute to meatpacking plants and industrial factories at nearby St. James, Jackson, Windom, Butterfield, Round Lake and Worthington.

Major places of employment for the new rural settlers include Hormel Foods at Austin, Seaboard Farms at Albert Lea, Monfort and Campbell Soup at Worthington, Caldwell Packing and Toro/Lawnboy at Windom, Swift– Echrich at St. James, IBP at Luverne, Jenny–O at Willmar, Ag–Chem Equipment at Jackson, Sanborn Manufacturing at Springfield, and Heartland Foods and Schwan's Sales Enterprises at Marshall, according to Minnesota Job Service offices in southern Minnesota.

Those jobs have attracted new Hispanic settlements to Willmar, Albert Lea, Marshall, Worthington and to communities nearby where housing was available. The Willmar immigration actually started more than a decade ago, predating the new wave of people moving to rural Minnesota, although the large Hispanic population there will be included with newer arrivals in future demographic data on the rural population.

Elsewhere, Vietnamese have settled in and around Austin; Hmong, primarily relocating from the Twin Cities, have moved to Lynd and Tracy near Marshall; and several hundred Somali and East African immigrants have moved to Marshall and nearby communities since 1991. Clem Schroeder, jobs service specialist with the Minnesota Job Service in Worthington, said Heartland Foods was expanding at the same time John Morrell cut meatpacking employment in Sioux Falls, prompting the Somali movement across the South Dakota border.

Paul Ehlers, manager of the Albert Lea and Austin Job Service offices, estimates that a third or more of recent hirings at meatpacking plants have involved minority workers who relocated to the area.

Austin and Albert Lea, however, each have about 20,000 people. The effect of the new immigration is more dramatic in small towns like Mountain Lake.

Teachers at Mountain Lake public schools, who once feared their schools would be consolidated into those of a larger community, now have full classrooms again. And 24 percent of their students are from Laotian or Hispanic households.

"A few years ago I was wondering where I might finish my career," said John Weir, Mountain Lake dean of students and athletic director. "I don't think about that now."

With 108 employees, the school system represents the largest payroll in town. By keeping the schools from reducing that employment, the new students have had an economic effect on the community, said Jim Brandt, director of the schools' English as a second language (ESL) programs.

The schools are working with the adult immigrant community as well, providing English language education in evening and summer sessions and helping new residents get their high school equivalency certificates.

State Job Service specialists at Austin, Worthington and Marshall say such educational opportunities are important. Some new arrivals don't have sufficient English skills to work at the Toro plant, for instance.

Lack of communication skills has contributed to some workplace problems, with new workers unable to articulate their needs so that managers can understand. In September, 80 Somali workers walked off their jobs at Heartland Foods in Marshall in a dispute that involved, among other issues, time for employees to leave the assembly line for Islamic prayers. Most eventually returned to work, and state employment specialists helped find jobs for the others.

Communication skills help smooth potential cultural clashes as well. "Let's face it, Mountain Lake is a pretty conservative town. Not everybody likes change," said Weir.

There have been no open breaches, and officials attribute much of the smooth transition to a multicultural community to the involvement of the churches.

Tong, the pastor, says "it means all seven churches joined hands with Lao Christian Fellowship to do the Lord's work."

Wilde, the economic development director, has a slightly different perspective. The seven churches dipped into congregational funds to bring Tong to the community, he notes – an investment that means everyone in town "has something riding on making this work."

Rural Minnesota communities increasingly will be challenged to follow the Mountain Lake model. State jobs specialists note that many rural workers moved away in the 1980s, and most southern Minnesota communities lost population. But in recent years, there has been a net gain in jobs, said Greg King, manager of the Toro plant in Windom.

"We've doubled our hourly work force since 1991, to 690 people here," King said. "We've been concerned about the labor supply because the seven counties around here have only 2.8 percent unemployment. That's so low we consider that to be full employment."

While Toro doubled, Monfort was expanding at Worthington. Now it is considering another meat line expansion that would create 150 more jobs. Immigration into the area has to fill the labor needs, said King.

Photo courtesy of Tom Sand

CHAPTER 8 THE NEED FOR A NEW APPROACH TO DEVELOPMENT

"Dr. Livingstone, I presume."
– Sir Henry Morton Stanley

"You see; but you do not observe."
– Sherlock Holmes

"Economic development is not the process of creating new facilities; it is the process of stimulating new activity – by individuals, by small companies, by large companies, by community development corporations, by chambers of commerce, even by local governments."
   David Osborne. Economic Competitiveness: The States Take the Lead. Economic Policy Institute, Washington, D.C. 1987.

Make no mistake. American agriculture has a bright future. World population continues to expand faster than world food production and global economic growth. Competitive uses for land keep shrinking the amount of acreage available for food production. Population pressures on land and water resources are creating environmental problems that limit food production in areas where it is most urgently needed. Some 1.1 billion people – nearly one–fifth of the world's population – were living in poverty in 1990. [8–1] They will either increase their food consumption by improving local production and from commercial imports, or the world community will respond to their plight with food assistance. Either way, North American, European, South American and Australian surplus food producers will gain markets.

"Today, there are more than 700 million people who do not have access to sufficient food to meet their needs for a healthy and productive life; they often go hungry and do not know when they will have their next meal," observed Per Pinstrup–Andersen, director-general of The International Food Policy Research Institute in early 1994. [8–2]

It "tarnishes" the world's image to have such huge numbers of people chronically hungry at the same time the world is considered to be food–secure because it does produce enough food, the research organization's director added. "Great progress has been made in meeting food needs during the last 30 years. For instance, the number of underfed people declined from an estimated 976 million in 1974–

197

1976 to 786 million in the late 1980s. But the problem is far from solved.

"Keeping up with increasing needs and demands due to population growth, income increases, and dietary changes is itself a formidable challenge." [8–3]

Given this outlook for world food needs, why should anyone be concerned about the future economic health of rural America and its communities? It seems rural America produces much to sate the world's needs and many of its wants. How could rural America fail in this economic and demographic environment?

Two reasons were discussed in earlier chapters. First, raw material producers – food, energy, minerals, forest products – are not insulated from government political and monetary interference in their markets, though the reasons for the interference may be accidental or coincidental. [See Chapter 3] Second, and what should be the more obvious to rural observers, high–tech food production no longer needs a large labor force or enough farm operators to sustain most rural communities from the revenues and jobs of retailing and marketing services. Therein lies the paradox: The future may indeed be good for American agriculture, but not for rural people. Taken a step farther, rural America could not tolerate another global feeding frenzy on American grains as experienced in the 1970s. Such strong export demand would lead to more consolidation of farmland into large operations and more investment in large scale equipment and technology, triggering an even more rapid exodus of people from the land.

But there is a third reason why some rural communities will fail in the years and decades ahead. It isn't as well understood and less easy to recognize. It affects communities that are making efforts to diversify their economies away from yesterday's labor–intensive agriculture. The cause of their demise will be the contents of the tool kit rural communities and state governments use to promote economic development.

Just as Explorers Stanley and Livingstone were drawn to the same Lake Victoria region of Africa at the same time, there are converging thoughts in the 1990s on the need to find appropriate methods for stimulating economic development. No clear consensus has evolved for future development actions; rather, appropriate methods are being arrived at by a process of elimination. This process leads to widely held beliefs that many past state and local attempts at economic development were counterproductive to community development. Oftentimes, communities were lucky when they didn't get the factories or other developments they sought. In the language of the countryside, they were chasing wild geese.

සාසාසා

This chapter will briefly review reasons why rural communities shouldn't count on the world need for food to sustain their economies. It will conclude with a discussion of economic development tools, describing what economists and planners sometimes refer to as the "three waves" of development programs. [8-4] The chapter's purpose is to set the stage for the following, final chapter, which will look at a new generation of cooperatives being formed that are consistent with current calls for a "fourth wave" of development action. As is true of the entire book, this chapter will draw from lessons learned by the author, other journalists and a number of social observers.

සාසාසා

The first lesson came from the Middle East. I arrived at Jeddah, Saudi Arabia, in early 1974 to interview Sheik Zaki Yamani, the Saudi oil minister. The Arab oil embargo was still causing gas lines and fuel shortages in the United States. The Royal Palace major domo, or chief of protocol, was apologetic and extremely helpful. He said Minister Yamani had been called to an emergency meeting in Vienna, Austria; but Prince Fahd, a half-brother to King Faisal who served as chairman of the policy-setting Saudi Arabia High Petroleum Committee, was willing to pinch hit for the oil minister.

The happenstance visit to Jeddah resulted in the future king of Saudi Arabia granting his first interview to a foreign journalist. [8-5] Equally unexpected, Prince Fahd gave oil politics only passing comment. He spoke instead of the oil economics and how Middle East oil producers share similar problems of markets with Middle American farmers. Run-away inflation in the United States, Japan and Western Europe was harming Saudi Arabia and other countries then involved in nation building, he said. OPEC, the oil marketing cartel, was raising oil prices in the early 1970s to keep pace with constantly higher prices its member countries paid for manufactured goods and imported services. The American media, he noted accurately, were ignoring the economic dimensions of OPEC and the Arab oil producers' dispute with the West.

"This is an important matter. If the United States, in cooperation with Japan and Western Europe, reach some agreement on the prices of

199

consumer products, we could lower prices," he said in the interview. [8–6]

The prince was well prepared for the interview. He knew which newspapers I wrote for, and that I had been writing from Washington about world food and agriculture policies. OPEC, he noted, was patterned after the International Wheat Organization the United States and Canada helped establish in London after World War II. The oil organization headquarters were established on the same London street as the wheat organization. [8–7] Moreover, he correctly observed, increased prices for grains, oil and other raw materials were not keeping pace with rising Western labor costs and manufacturing inflation; but governments throughout the developed world were blaming OPEC and North American farmers for causing inflation problems.

Hardly a month has passed in the 20 years since when I haven't thought about King Fahd's comments, and how they've been supported by subsequent world events. Inflation is an economic concept rooted in imperial Europe. It is a valid and accurate economic measurement, certainly. But it is a concept developed by manufacturing nations that relied on inexpensive raw materials from at home and abroad, and cheap labor to extract commodities and load them on ships. Imports from colonies allowed the citizenry and businesses in industrial countries to enjoy the benefits of colonial resources, Prince Fahd said.

This was also the American colonies' experience before independence, he noted. It troubled him that nothing was built into economic theories to recognize intrinsic values of nonrenewable and renewable resources. Though I am now paraphrasing the king from memory, [8–8] he said oil producing nations were particularly upset that people in Western countries failed to recognize OPEC pricing as a response – not the driving force behind Western inflation. For instance, Americans didn't grasp that Western oil refiners' price increases and Western taxes on oil products raised consumer prices more than OPEC price hikes.

Within a month after I returned to Washington and resumed writing about food trade issues, Kansas' Senator Bob Dole and Congressman Keith Sebelius both vented frustrations with the Japanese in news articles our bureau wrote for the Wichita Eagle newspaper. The Japanese were complaining about the escalating cost of American farm products, especially soybeans that had more than tripled in price between 1972 and 1974. President Ford soon thereafter became president and embargoed soybean sales to Japan to fight domestic inflation. The two Kansas lawmakers, however, noted Japan's tariffs on American soybeans raised more money for Japan's treasury than

American farmers were being paid for the beans. If the price was getting too high for Japan's economy, they argued, then Japan should reduce its tariffs to lower costs for its soybean users and consumers. [8–9]

It was the same argument I heard earlier in the Middle East.

ဢၵၢၵ

A second lesson soon revealed itself after the Arab oil embargo. For any number of economic reasons, industrialized countries will cannibalize raw materials, including their own.

Presidents and members of Congress in the United States went through blame–throwing exercises over fiscal responsibility, as did prime ministers and parliamentary opposition leaders in other Western democracies. Short–sighted fiscal policies contributed to problems that fell from the powerful, developed countries onto the world economy. But a radical change in world monetary policies, as noted in Chapters 1 and 3, had an even greater impact on the flow of trade, foreign exchange rates, world commodity markets and the health of the U.S. economy.

The inflation of that earlier era, and variations that were sometimes called stagflation, prompted President Richard Nixon to try wage and price controls to attack inflationary expectations. President Gerald Ford tried to "jawbone" inflation out of existence with a WIN ("Whip Inflation Now") campaign and an embargo on soybean exports. Running out of options, President Jimmy Carter gave the international finance community what it wanted – a club–wielding monetary policy chieftain who would take no prisoners in his war on rising prices. All three presidents worked to mitigate the domestic economic consequences of the greatest U.S. currency devaluation since the American Civil War. [See Chapter Note A, Chapter 3] President Ronald Reagan took political credit for finally bringing inflation under control, although it was a gift handed him by the Federal Reserve Board. [8–10]

President Nixon's wage and price controls hadn't worked and were abandoned. They did slow the economy, but they only put inflationary pressures on ice, waiting to be revived once the controls were lifted. President Ford's WIN campaign didn't work either, and his soybean embargo only lowered farm prices for soybeans and gave minor relief to domestic soybean processors. There were doubters aplenty heading into both inflation fighting programs, led by economists. But I won't criticize either attempt; had they worked, they would have avoided the far more painful option that was left for President Carter. That option was an inflation–choking, tight monetary policy unleashed by Fed

Chairman Paul Volcker. It promptly threw the national economy into a wringer that did squeeze inflation out of the system. The high interest rates and tight money effectively sacrificed the farm, forest, mining and domestic energy industries, and the rural towns that served them, while value–added manufacturing and service sectors weathered a milder, two–year recession during a period of national economic adjustment. [8–11]

I don't expect all readers to understand or agree with the political and economic descriptions offered above. Americans like to debate fiscal policy, not monetary policy. They give presidents too much credit and blame for economic events, not appreciating limits on presidential power or economic trends that run cycles indifferent to four–year terms of office. But these experiences from the recent past should be remembered by farmers and community leaders preparing for the Twenty–First Century. In early 1994, the Federal Reserve Board was again raising interest rates. The U.S. economy was heating up. Unemployment was falling to relatively low levels. There were signs that inflation could return with greater consumer spending. Hence, the hammer.

The Fed of the mid–1990s, being led by Alan Greenspan, was squeezing credit to slow the economy. The stock and bond markets reacted first with large investor selloffs. At this writing, it wasn't possible to predict how far the Fed would go in raising interest rates to keep the U.S. economy from overheating. The U.S. dollar was falling in value against the Japanese yen, but not against other trading partners' currencies. It wasn't possible to gauge what the trade response would be. But one response is predictable from past experiences: If a tighter monetary policy is stretched over time to inhibit inflation, the lower rungs of the industrial ladder – basic raw materials and basic labor – will be again clobbered as they serve as shock absorbers for the broader economy.

৪৩৪৩৪৩

Looking back, Historian Emily Rosenberg has concisely assessed the past century of development in the Northwest quadrant of the United States:

"Ironically, the periods that brought greatest prosperity to the area's extractive and agricultural pursuits also, by entrenching greater dependence on natural resources and straining the environment, foreshadowed grave problems for the future. While the revolutions in production and transportation brought settlers and created livelihoods, resource dependence created many problems commonly associated today with Third World extractive economies: lack of local control, boom and bust cycles, environmental stress." [8–12]

ഇഇ

That raises what should be a third lesson. All across America, rural communities and political leaders are willing to trade away strengths to compensate for weaknesses. Where rural communities and their political leaders perceive the need to move their economies to higher rungs on the economic ladder, they also have turned to development strategies similar to those used in Third World countries.

The most basic form is what the Council for Enterprise Development calls a "first wave" strategy, and what casual people call smokestack chasing. To diversify from a dependence on nearby resources, these communities give away tax obligations that would pay for local services, and substitute a new dependence on jobs that demand inexpensive labor. The only way these jobs are secure is if labor remains inexpensive. For starters, factories that need inexpensive labor to remain competitive are the first to pack up and move off shore when faced with low–cost competition from abroad. Jobs are an easily transported commodity.

A community's new dependence on inexpensive labor and tax subsidized workplaces creates chain reactions that further harm the community commonweal. The low–paid work force needs local services, from education and welfare to increased police and fire protection. Low–paid workers can rarely pay for services with their tax contributions. To compensate, communities must increase the tax burden and seek other subsidies from traditionally stronger sectors of their population. In Egypt and many countries of Africa, the stronger sector usually targeted to provide income transfers is agriculture. This, in turn, weakens the stronger sector and discourages investments needed to improve agriculture and the nations' integrated food systems. South Dakota, Mississippi and Texas, to name just three American domains, do the same – letting property taxes on farms and resources substitute for income taxes and business taxes. So far, these states with regressive tax structures have avoided damage to their agriculture similar to that experienced in many Third World countries.

I was visiting in Cairo when the late Anwar Sadat Egyptian President was nearly toppled from power for trying to remove price controls on agriculture. Egypt was subsidizing food consumption for the masses by keeping lids on farm and food prices. Egyptian economists and consultants from international agencies were telling President Sadat that Egypt should subsidize consumption and combat poverty with

203

income transfer programs, such as America's food stamps. This would allow Egypt's agriculture to expand and ignite a new round of economic development in communities up and down the Nile. The people preferred cheap food to economic arguments and development theory. Sadly, I must report, Egyptian President Mubarek is still struggling with the now–institutionalized cheap food policies.

I wasn't on the scene, but I watched from America as a Communist regime in Poland was nearly toppled for the same reasons. This was before Lech Walesa and the Solidarity Movement arose and did bring down Poland's Communist government. [8–13] It is fitting, perhaps, that Poland under Lech Walesa and Western–style democrats has allowed food prices to rise and market forces to return to agriculture even though similar experiments caused great political problems for their predecessors.

What people in Egypt, Poland and throughout America all tend to forget is that old Midwestern proverb: There is no such thing as a free lunch.

ഇരുഇരുഇരു

Dan Chapman, one of America's outstanding business writers who keeps a close eye on economic development issues, has chronicled some of the biggest plays communities have made to attract factories with first wave development incentives. He writes for the Charlotte <u>Observer</u> in North Carolina. Community and state economic development projects make daily news stories in his rapidly growing area.

North Carolina lost one of the biggest smokestacks it ever chased, and Mr. Chapman, for one, thinks the state may be fortunate. At stake was the $300 million Mercedes–Benz sport–utility truck plant that the automaker chose in 1993 to build at Vance, Alabama. He said in an interview shortly after Mercedes made its decision that a number of state officials weren't too unhappy, either. "The stakes were getting too high. You couldn't tell if the pay–back would equal the pay–out." And it wasn't just government units that were opening doors and pockets to the German firm, as he reported in the following analysis story:

Behind the Bid for Mercedes
State Benz over backward in failed effort, files show

By Dan Chapman, Staff Writer
RALEIGH – North Carolina unlocked its Mercedes–Benz trunk. Look what came tumbling out:

IBM Corp. tried to drive Mercedes home to Charlotte.

USAir offered hundreds of free tickets to the Germans.

The governor was practically ordered to say nice things about Mercedes when Alabama won.

A private White House tour was dangled in front of a Mercedes executive.

The reams of confidential memos, environmental findings, testimonial–like letters, unsigned contracts, handwritten notes, maps, videos and other recruitment paraphernalia offer a little–publicized insight into the hunt for the year's most coveted industrial prize: the Mercedes sport–utility factory.

The files – made public recently by the N.C. Commerce Department – shed light on the secretive site–selection process, code–named Project Rosewood. They also offer an intriguing – and sometimes silly – look at how the Carolinas' titans of business and government tried to lure the Germans here.

All for naught. Mercedes surprised the Carolinas and just about everybody else when it announced Sept. 30, (1993) that Vance, Ala., would be the sweet home for its $300 million factory and 1,500 jobs. Privately, Gov. Hunt, N.C. Commerce Secretary David Philips and dozens of state and local business recruiters were dumbfounded and angry that they had lost Mercedes.

Publicly, they were gracious losers.

A peek into the files explains why. They were urged by Mercedes to be nice. Just three weeks before Mercedes made its decision public, its lawyers worked up a contract saying if North Carolina was not chosen, Tar Heel state officials must still 'respond in a positive light.' "

Hunt and Philips were gracious in defeat. Nonetheless, they said North Carolina lost because the Tar Heel State couldn't match Alabama's $300–plus million in tax breaks. All along, however, Mercedes said money wouldn't be the deciding factor. Maybe the Germans didn't tell Fluor Daniel Corp., the company that handled the search for Mercedes' site.

In a confidential May 3 letter to Watts Carr, the commerce department's top recruiter, Fluor Daniel wrote that recruitment incentives "must be very attractive to be considered further."

The commerce department started its file in March 1990 when the automaker was reported to be interested in building some type of vehicle in the United States. North Carolina was considered the front–runner because Freightliner owned by Daimler–Benze AG (Mercedes) Corp., owned by Daimler–Benze AG, already had three factories making trucks near Charlotte. Just about everybody in North Carolina was comfortable wearing the front–runner mantle.

Except Carr.

In a cryptic and somewhat pessimistic memo penned May 5, 1993, Carr wrote that "Mercedes wants own culture, no real synergy with other operations expected." He also lamented that neither of North Carolina's two main ports, Wilmington and Morehead City, was big enough to handle Mercedes.

Here are some other behind–the–scenes tidbits culled from the state's files:

* **IBM Corp.** offered Mercedes 600 acres at its Charlotte campus along I–85 and Harris Boulevard. When informed that was insufficient for the German automaker's big–time industrial needs, the computer giant upped its ante to 856 acres, according to a July 13 letter. When told Mebane in Alamance County was the top N.C. choice, IBM sought the $35 million automotive engineering school, dubbed Mercedes University. Persistence didn't pay.

* **NationsBank,** First Citizens Bank and Wachovia were willing to put their money down to sell Mercedes on the Carolinas.

Charlotte–based NationsBank, for example, offered to lend hundreds of millions of dollars at generous rates to Mercedes to build the plant. The offer was first made in North Carolina, but after S.C. Gov. Carroll Campbell got wind of it, NationsBank extended its generosity south of the border.

First Citizens of Raleigh would have provided cut–rate mortgages to 50 Mercedes executives if they banked with them. Other banks reportedly offered similar plans.

* **Charlotte–based Duke Power** would have helped buy the land in Alamance County for Mercedes. Duke took out options on the land and was ready to build a substation solely for Mercedes.

Chairman Bill Lee went a step further, however. On June 24, he wrote letters to a German senator, the chairman of a German metalworking company and a German vice chairman with J.P. Morgan asking for help in bringing Mercedes to North Carolina. Lee wanted the gentlemen, in their discussions with Mercedes officials, to find out what Mercedes was thinking about North Carolina and what suggestions they had for strengthening the N.C. bid.

Lee's letters weren't all devoted to industrial insider information. He inquired as to the condition of Sen. Hermann Becker's back. He also asked to be remembered to the senator's wife.

Duke was asked at other times to use industry back channels to provide the commerce department with information. Secretary Philips wanted to know whether the federal Tennessee Valley Authority could turn over land to Mercedes in Oak Ridge (Tenn.), according to a Sept. 7 letter. Duke's Jack Roddey determined the TVA could legally transfer the property, much to the chagrin of the commerce department.

* **The airlines** chipped in, too. USAir and American Airlines offered $500,000 in free tickets and other gifts to Mercedes.

USAir was willing to provide 100 economy class transatlantic round–trip tickets, 100 domestic round–trippers, 25 tickets for charity, 25 for employee incentives and 25 USAir Club memberships. All free.

That's not all. The airline offered four round–trippers from Germany to North Carolina for two top Mercedes executives and their wives for a final site tour in September. And USAir said it would publish a puff piece in its in–flight magazine about Mercedes.

What did Mercedes need to do in return?

Designate USAir as Mercedes' "preferred supplier" for air travel in the United States, according to a Sept. 17 letter. Oh yeah, and build its plant in North Carolina.

Not to be outdone, American Airlines offered 100 free round–trip tickets from Raleigh–Durham International Airport to Germany and to establish an ongoing "International Incentives Program" – frequent flier–type program, according to a July 20 letter to Secretary Philips. [8–14]

ഇ ഇ ഇ

All states, communities and area businesses engage in smokestack chasing. What makes Mr. Chapman's account so interesting is that he was able to reveal far more of a "first wave" package than what usually comes to light. What's more, the North Carolina give–a–ways proved to be no match for Alabama's offer. Though details of the winning bid aren't as well known, Alabama did offer more than $300 million in tax breaks for Mercedes to build a $300 million plant. [8– 15] Hopefully, Alabama has done careful cost accounting and anticipates that Mercedes will pay its 1,500 workers sufficiently high salaries to repay the state through taxes and new economic activity. That won't happen if Mercedes' decision was partly based on expectations of low labor costs despite a North American site. But assuming the plant will pay prevailing automaking wages and salaries, the payback for state and local services will be gradual. Who pays the freight in the interim? School children attending northern Alabama schools? Area farmers and businesses, through higher tax rates on their properties and incomes?

A number of Midwest and Southern states fashioned similar first wave incentives to lure the Saturn auto plant that was built in Tennessee. In another recent example, the Minnesota Legislature passed legislation in 1994 for the city of St. Cloud to extend its 10–year first wave tax breaks to 25 years, under certain circumstances. The circumstances were better known as Fingerhut Companies, the giant mail order and direct marketing concern that is now expanding at three sites in Minnesota. [8–16]

When such strategies do produce results that include new jobs and new economic activity, it isn't easy to criticize, says Richard Mattoon, regional economist with the Federal Reserve Bank of Chicago. "We (economists, planners) can get together at conferences and criticize give–a–ways of land and taxes by communities chasing factories. I do it myself. But I've heard good arguments from people at the Indiana Economic Development Council who say it isn't all bad. They remind us that some communities don't have anything else to offer. They are like Third World countries. They have to get economic development going, so they use the only tools they have to get it started." [8–17]

Economist Mattoon and his colleague William A. Testa have published several articles on development theory and state practices to help the Seventh Federal Reserve District states of Illinois, Indiana, Ohio, Michigan and Wisconsin plan their futures. These states all have a large industrial base from which to expand, unlike many of the more resource–dependent states of the Northwest. But these diversified Midwest states also employ elements of all three waves of development strategies. They still bring out big guns in showdowns over major auto and other manufacturing plants, the economists note; but the primary era of the first wave was the 1960s and 1970s.

Taking a cue from a popular Hollywood movie, the state of Tennessee sent "Ten Tall Men from Tennessee" around the country to promote the Volunteer State as a site for wandering industry and factories. On one such foray into Minnesota, Gov. Harold Levander turned for help from former Minneapolis Lakers center and NBA Hall of Fame legend George Miken. When the 10 tall Tennesseans arrived at the airport, 10 taller Minnesotans were there to greet them. Take a good look around, the visitors were told. They'd be welcome if they decided "to defect." [8–18].

80:80:80

The folly and expense of chasing smokestacks became apparent to state and community planners with the opening of international markets for most goods and services. You can sell yourself cheap, but you can't keep rootless factories at home if people in other states or countries are willing to sell themselves or their environment even more cheaply. A second wave strategy was inevitable.

80:80:80

"In the second wave, states tried to help firms compete in the global economy and to encourage entrepreneurship," wrote Economists Mattoon and Testa in summarizing discussions at Great Lakes development conference. [8–19] "This approach led to a wide range of development programs, from financing and export initiatives to technology transfers. While these programs seemed promising, they were often fragmented and uncoordinated and therefore often did not produce the intended results. Lacking both accountability and scale, these efforts failed to leverage private sector resources." [8–20]

States and communities have found clever second wave operations that do help their economies. And they have second wave instruments in place that can't be measured for success or failure, although they often are graded by politicians and the public without the use of valid measuring tools.

The state of Arkansas and several major American cities have established small offices that have proven to be positive, though still hard to measure, second wave projects. These are the offices established to provide services for the film industry. These small operations encourage film makers to use their towns or natural scenery as on–location sets. The money spent locally by the studios for services, supplies and entertainment give a spurt to the local economy and more than pay for these one–, two–and three–person offices.

Economist Wilbur Maki at the University of Minnesota has created econometric models for estimating the economic benefits derived from some second wave projects. The state of Oregon has gone even farther and established a benchmarking system for evaluating many of its state programs in what is proving to be both an admired and copied system for other states. But how does any state measure the impact of a state trade office? Almost all states have a major trade office and the larger economy states have trade centers in addition to trade offices. The up front purpose for most of the Midwest state trade offices was to help small and medium–size companies grow by entering international markets. In Iowa and Minnesota, for instance, there was no delusion that state offices would help Pioneer Hybrid International or Cargill Inc. with exports – both were masters of their respective international markets. What's more, says Michael Bonsignore, chairman and chief executive of Honeywell Inc., human nature often prevents trade offices from receiving credit for business deals when credit is due. What vice president for international sales is going to step forward and share that credit, he asks. [8–21]

Any state or local agency created to assist businesses and communities with economic development becomes an easy target for political attacks. This is especially true for people whose politics are locked in Twentieth Century class struggles. Depending on the politics of the governor or mayor, critics will see any second wave effort as a boondoggle for business or spoils for "hacks." But there are legitimate criticisms of second wave programs, observes Dan Pilcher, an economic development specialist with the National Council of State Legislatures. In 1991, he noted, these programs assisted only a small percentage of firms in a state or region. And most state programs aren't provided with enough funds to be effective. [8–22]

ଊଊଊ

The "third wave" was spotted by the Corporation for Enterprise Development in 1990, and it has spread across the country. As the development specialist Pilcher sees it:

"The major strategies of the third wave are all long–term investments. Investments in people through education and workforce skills. Investments in distressed communities through help for the people and businesses in those communities. Investments in programs that encourage business and local government to work together." [8–23]

Models for "third wave" policies were found in the regions and provinces of Baden–Wurttemberg in Germany, Jutland in Denmark, Smaaland in Sweden and Emilia Romagna in Italy. Washington and

209

Oregon were among the first states to send researchers abroad to study economic development successes in those countries. [8–24] What they found were local and regional assistance programs, including investment funds, for industries.

Companies within the targeted industries work cooperatively together, aided by infrastructure investments such as education and training. If North American comparisons can be cited, it would include Silicon Valley in California, Medical Alley in Minnesota and Manitoba, the insurance centers of Connecticut, and a growing number of new manufacturing centers that share market information and in some cases, equipment. Among such centers cited by the state legislatures' organization are the Manufacturing Innovation Networks created in Pennsylvania, flexible manufacturing centers in Ohio, metal working in Arkansas and small company defense contracting in Montana. [8–25] [8–26]

The Chicago Fed's Mattoon and Testa add the Michigan Strategic Fund for development financing, Edison Technology Centers in Ohio, and the Ben Franklin Partnership and Industrial Resource Centers in Pennsylvania as examples of the new development principles. But they also note that waves follow waves:

"While these (above programs) are promising efforts, it is still not clear whether programs of this scale can improve the competitive position of the nation. Perhaps a next wave will involve regional cooperation to leverage greater amounts of resources." [8–27]

ഇന്ദ്രഇന്ദ്ര

Perhaps.

If there is a lesson to be learned from state and local efforts at economic development, it should be this: Almost all such efforts are attempts to change business environments and economic trends to justify the presence of industries or companies where they didn't develop naturally. Efforts to totally bend economic forces are likely to fail. But attempts to develop community resources – including local leadership – gives economic development a fighting chance.

"Creating new community programs is a way of mobilizing local leaders," says Wilbur Maki, the University of Minnesota economist who specializes in development issues. [See Chapter Note A.] "These leaders begin to see that change is possible and government exists to support change. They see possibilities of networking and linkage with organizations in other areas – the metropolitan core area, surrounding labor market areas, and even more distant area service centers." [8–28]

Third wave or emerging "fourth wave" development programs aimed at helping local efforts at development serve as demonstrations that show local leaders what they can achieve, he says. "This is a far cry from depending on the federal government and its distant offices that cannot treat one region differently from another." [8–29]

Communities, Professor Maki notes, have changed and no longer reflect the political divisions of earlier times. [8–30] They have become the business and service areas that surround core cities that house major airports, be they the nation's international airports or the feeder airports that tie into the international depots. Smaller communities may well involve all of a county or connected portions of several counties. Planning should involve the whole community, even when it is divided by state lines or other political boundaries.

Taken out to a logical conclusion, Iowa, Minnesota, South Dakota and Nebraska should have rural economic development policies that recognize the shared livestock market region served by John Morrell in Sioux Falls, South Dakota, and IBP at Dakota City, Nebraska; as well as other meatpackers in the four corners of those states. And Wisconsin and Minnesota, which already cooperate by accepting each other's students at in–state rates of college and university tuition, should cooperate more so western Wisconsin communities may participate more fully with the Minneapolis–St. Paul core metro area to which they are economically bound. Northern South Carolina has a similar linkage with Charlotte, North Carolina, as noted earlier. Similar linkages exist with Iowa communities near Omaha, Nebraska; Kansas suburbs and exurban cities near Kansas City, Missouri; the Quad Cities of Illinois and Iowa; Fargo–Moorhead, with an economic influence in three states along the Red River of the North; Cincinnati and northern Kentucky, and countless other areas across the nation.

The biggest example – and perhaps best example – of shared community is the seaport, airport and trade complex of New York and New Jersey that works as one entity. But even as the international trade and transportation complex makes New York City a full–fledged partner with the states of New York and New Jersey and adjacent New Jersey municipalities, the city has embarked on its own third wave development strategy aimed at luring companies to relocate in the city. [8–31] The New York Times quotes the city's economic development director as saying New York "is open for business again." [8–32] No offense to the Big Apple's leadership, but if city officials want to understand third wave development programs that work, they could go to lunch atop the New York World Trade Center Towers. There, beneath them, they will see the cooperative New York–New Jersey port complex and the trade offices they attract. This two–state trade, finance and transportation complex will remain the best reason for businesses to locate in or near New York.

Development programs that recognize shared interests of these new economically linked communities will work at building on the enlarged community's strengths. In rural America, that includes support of infrastructure to help develop projects such as value–added processing of locally produced commodities and area resources. They offer the best hope for stimulating economic development by individuals, small companies, large companies, community development corporations, chambers of commerce – as was suggested earlier. The next chapter will look at how a new generation of cooperative businesses – community, producer, consumer and joint venture cooperatives – can be rural America's best new wave vehicle for economic development.

This new wave, or "fourth wave", is starting to make a splash.

Chapter 8 FOOTNOTES

1. International development agencies are in general agreement on the number of people living in poverty, although the exact world population isn't known. The United Nations Food and Agriculture Organization (Rome), the Organization for Economic Cooperation and Development (Paris), and The World Bank (Washington, D.C.) use the 1.1 billion poverty population estimate, based on 1990 figures.

2. Pinstrup–Andersen, Per. "World Food Trends and Future Food Security." (Published monograph) The International Food Policy Research Institute. Washington, D.C. March, 1994.

3. Ibid.

4. "Waves" of development programs are the theoretical invention of the Corporation for Enterprise Development, Washington, D.C., and are occasionally credited to both Brian Dabson and Bob Friedman. They will be explained later in the chapter.

5. The interview with Prince Fahd, who would become king later, appeared in Ridder Publication newspapers in January, 1974. One account, "Prince Ties Oil Tab, U.S. Profits," appeared in the Pasadena (California) Star–News. Jan. 13, 1974.

6. Ibid.

7. The International Wheat Organization was established in the early 1950s to work as a cartel–like wheat trade organization. There was never the political support within its member countries, however, to use cartel powers to influence world grain prices. It has functioned as an information sharing organization for its members by gathering data on world food cereal grain production and anticipated needs. Daniel Amstutz, the former Cargill Inc. executive who was the principal trade negotiator calling for an end to all farm and trade subsidies in the early Uruguay Round trade negotiations, became director of the IWO in the 1990s. OPEC, meanwhile, borrowed the IWO charter and did become a cartel organization in 1955.

8. It has been too many years since the interview to attempt to quote King Fahd precisely, though I have since discussed the contents of his remarks with economists in the United States and Europe. The only disagreement with the king's comments on economics, and it is slight, is that raw materials would have intrinsic value in a "pure" market in which all elements of production and consumption are allowed to find their value. As long as there are tariffs and domestic taxes applied to raw materials and finished products, such a pure market doesn't exist.

9. The argument raised by Kansas' Sen. Bob Dole and the late Rep. Keith Sebelius is identical to the complaints of OPEC as expressed by King Fahd. Nearly all farm trade and promotion groups, including the American Soybean Association, have complained about Japanese tariff barriers to American farm exports in the past two decades.

10. The author's observation. The tight money supply and high interest rates ushered in by Fed Chairman Paul Volcker discouraged corporate and individual borrowing and use of credit. It raised the value of the U.S. dollar above comparable rates for other hard currencies, thus discouraging U.S. exports and manufacturing while stimulating U.S. imports. It ended America's problems with inflation while creating balance of trade problems; and it continued balance of payments problems for the United States. It also deflated America's raw material economy, including agriculture, which suffered a five–year depression while the broader U.S. economy endured a 1980–1982 recession.

11. It's a challenge, but the author wants to believe the lingering political attacks on President Carter for causing inflation, high interest rates and related economic problems come from innocent ignorance, not demagogic politics.

12. Rosenberg, Emily. "The New, New West: Assessing a Century of Change in the Northwest Region." (Published essay.) Northwest Area Foundation. St. Paul, Minnesota. 1991. Ms. Rosenberg, who grew up in Montana and Wyoming, is a professor of history at Macalester College, St. Paul, Minnesota.

13. Only farmers and serious international affairs watchers may remember this. Poland was a major market for American corn. In the mid–1970s, American farmers were hopeful that economic reforms in Poland would lead to expanding livestock production, which in turn would create more demand for corn. Internationalists were watching to see if the Polish Communists' experiments with market forces would move the country away from Soviet dogma. Food riots brought the experiment to a quick end, and the food rioters were soon replaced in the streets by Poles seeking political freedom.

14. Chapman, Dan. "Behind the bid for Mercedes." Charlotte Observer. Oct. 11, 1993.

15. Chapman. From an interview in November, 1993. 16. From news accounts, St. Paul Pioneer Press, and Star Tribune, Minneapolis. 1994.

17. Mattoon, Richard H. From telephone interview conducted August, 1993. Mattoon is regional economist, Federal Reserve Bank of Chicago.

18. As recalled by Bob Swanson, a great teller of Minnesota lore and long–time spokesman for the Minnesota Department of Agriculture who had served as press secretary to Governor Levander.

19. Mattoon, Richard, and Testa, William. "Shaping the Great Lakes economy: a conference summary." Economic Perspectives. Federal Reserve Bank of Chicago. July/August, 1993.

20. Ibid.

21. From comments at a trade conference at the Minnesota World Trade Center, 1989.

22. Pilcher, Dan. "The Third Wave of Economic Development." State Legislatures. National Council of State Legislatures. November, 1991.

23. Ibid.

24. "Oregon Looks to Europe for a Model." State Legislatures. National Council of State Legislatures. November, 1991.

25. It should be noted that state trade offices, created under second wave development strategies, now serve third wave purposes of providing companies and industries with important market information.

26. State assistance to industry networks, such as described in the preceding paragraph, also resemble government sanctioned and supported cooperation within Japanese industries. A slightly different structure, but with the same purpose, is the collaboration of Dutch government research, targeted extension education and business cooperation described in Chapter 4.

27. Mattoon and Testa. Economic Perspectives.

28. Maki, Wilbur. "When is a Government Loan to a Fortune 500 Business State Economic Development?" A paper published by the University of Minnesota. 1993.

29. Ibid.

30. Professor Maki outlined his definitions of new communities being shaped around international and regional airports at a meeting of the National Association of Agricultural Journalists in October, 1991, at Minneapolis, Minnesota.

31. Redburn, Tom. "New York Pressing Bid to Lure Companies." The New York Times. New York, N.Y. . June 6, 1994.

32. Ibid.

Chapter 8 CHAPTER NOTE 8–A

Dr. Wilbur Maki, a professor of agricultural and applied economics at the University of Minnesota, provided a wealth of information on community economic development tools and concepts after critiquing this chapter. I have included some of the material above, and have inserted comments and observations from his friend, Canadian economic development leader Bill Hatton, in the next chapter.

But I haven't included all of Professor Maki's comments or his offered materials. This book is written for people in rural communities from farm, business and civic backgrounds who are interested in reviving their communities. This chapter wades in on serious academic debate over development policies, but my purpose was to serve as a bridge between scholars and the community people and their leaders who will ultimately make development decisions.

For those community leaders, however, who may want to consider development issues more deeply, I offer the following excerpts from Professor Maki's comments:

1. *How can rural America fail?* By accepting the belief that modern, technologically–advanced economies like ours can survive without primary resource production while at the same time allowing developing countries to use their primary resources for international trade.

This means, if our beliefs are consistent with reality, that we can close our national forests and all ecologically vulnerable agricultural lands without adverse repercussions on the rest of the economy. This is not a matter of rational choice, but an integral part of an emerging system of beliefs about the primacy of achieving an ecological condition from which emanates man's most profound spirituality – whatever that means.

(At the same time,) the good news associated with high–tech food production is that it helped set up the rural infrastructure for the new manufacturing in rural areas within 75 to 100 miles of core metropolitan areas.

2. *Lessons learned.* First, developing countries – like many rural areas – produce a standardized product that is traded worldwide. Success depends on constantly reducing production and transfer costs Developed countries with emerging information–based economies depend on services that are not standardized and, hence, do not compete on a strict cost–of–production basis.

Second, developed countries with decentralized markets and populist civil governments heavily subsidize consumption . This imposes a double burden on the developed economies by reducing the social surplus for investment and adding to the hidden costs of environmental pollution.

Third, the service sector in developed countries now accounts for 70 percent or more of the total economy. It enjoys some of the benefits of a spatial monopoly. This is particularly the case for the health services sector that has the added attribute of being financed by third parties, not the user of the services. This contributes to runaway costs and inflation that trigger the boom and bust cycles of the commodity markets and fuel the continuing conflict between developed and developing countries. US inflation contributed to much speculative investment in which many US farmers participated, some to their ever–lasting sorrow. The cure through high short–term interest rates devastated the interest–sensitive sectors. This cure also left a legacy of high long term interest rates that discouraged many output–increasing investments in private plants, equipment and public infrastructure.

Fourth, the higher incomes of developed country residents supports ambitious social programs that provide a "safety net" for unemployed workers. These programs have devastating consequences for the fiscal health of both developed and developing countries. They push up the acceptable full employment – unemployment level from 4 percent to 6 percent or more, while adding to the public cost of maintaining the benefits and at the same time reducing the overall productivity of the economy. This means there is little public surplus left to assist developing countries to build their own economies. (As with issues cited above, this also contributes to weakening trading partners and reduces the cost of standard commodities sold by developing countries.)

Fifth, the growth and power of the military–industrial complex, fueled by the Cold War hysteria, added further to the fiscal burdens of civil governments. It probably prolonged the Cold War by a decade or more at a cost to the United States alone of more than $3 trillion. We now wonder why we are unable to help the developing countries to help themselves.

3. *Lessons on the "waves."* The progression of economic development strategy from chasing smokestacks to helping firms compete in global markets on up to making investments in education and training and encouraging business and communities to work together is like moving up a learning curve on a product cycle.

The first wave was the backside of the product cycle nearing the end of its useful life. It was all down from there, with rural areas being the burial ground for many businesses that could no longer compete, except by continuing to drive down production costs.

We found out belatedly that we had a lot to learn about the product cycle that has a pre–market gestation period, an early market entry period of rapidly expanding sales and profits, a period of approaching maturity with tough competition from many producers, and finally, a period of declining profits and sales.

Entrepreneurship seemed like the secret to product cycle success, but entrepreneurship with the capacity to compete globally. It was soon evident that entrepreneurs must have skilled workers for producing a marketable product and the regional infrastructure for reaching markets where the product can compete successfully. Still nothing happened.

What's missing? Management! Entrepreneurial management! This means having the business enterprise select the right product to sell in the right markets at the right price. These are learned skills with the end–in–view of creating a successful enterprise that brings new dollars into the community and earns a reasonable profit on its investment. This makes the business enterprise an integral part of a viable local industry cluster that is the outcome of successful community–based economic development.

4. *Models for estimating benefits from economic development projects.* With second wave economic development being an elusive concept, the models for measuring results were more qualitative than quantitative. My current modeling efforts are directed toward fourth wave economic development with emphasis on regional exports and imports and interregional trade – its origins and destinations by detailed commodity categories.

We now have the beginnings of the sort of tools community–based economic development managers seek in making their performance–based investment decisions. The key difference between community–based economic development and community development focuses on the purpose of the development activity. Is the purpose to create jobs or to create products that can generate profits for sustaining the continuity of the enterprise? Jobs must become secondary to producing a product that competes successfully in the export markets and thus brings new dollars into the community.

CHAPTER 9 COOPERATIVES:
THE NEW WAVE MOVEMENT
FOR COMMUNITY DEVELOPMENT

"Economic development strategies have too often been narrowly focused, such as on jobs! If the emphasis is upon community development, then all the interests of the community are represented – not just a business' or businesses'.

"Cooperatives are uniquely positioned to lead community development. By their nature, they have an inherent interest in sustaining community. This fourth wave of economic development is best lead by co–ops."

Gary DeCramer, fellow of the State and Local Policy Center, Hubert H. Humphrey Institute of Public Affairs, University of Minnesota.

A new rural America is starting to take shape, rising from the ruins of communities that were no longer needed to serve the needs of traditional agriculture. It is rising up in sparsely populated counties of the Dakotas that have been in steady decline since the 1870 Census, as Professor Randy Cantrell noted at the beginning of this book. [9–1] It is evolving in rural communities of Wisconsin, Idaho and Wyoming where local cooperatives, working with large regional co–ops, have expanded community services and diversified businesses on a horizontal path of expansion, or integration. More recently, since about 1990, it is happening in Iowa, Minnesota and North Dakota communities willing to jump start the development process.

Where this is happening community residents are building new generation cooperatives that follow both horizontal and vertical paths of expansion – horizontally reaching out to produce new crops and products, bringing diversification; and vertically, from the ground up, while creating jobs for rural townspeople and giving area farmers some of the profits from processing their commodities.

Some of this transition to a new rural America can be traced to creative community responses to the farm financial crisis of the 1980s. Town and country residents, using their local co–ops, reached out and took over Main Street retail businesses and local services that were collapsing. More recent developments, however, have come in more prosperous times for agriculture. Farmers and townspeople have assessed the changes occurring in farming and world markets and concluded that they, too, must change.

In 1989, for example, Professors Frank and Deborah Popper from Rutgers University in New Jersey infuriated a large number of people in the West. They wrote an essay, **Buffalo Commons,** that was actually a scholarly look at sparsely populated areas of the West and Great Plains and concluded that farm programs and other federal assistance to the West was a terribly expensive, ineffective way to operate a regional welfare system. [9–2] In effect, the Rutgers academics used data to argue that federal payments into the low–population counties of the West exceeded any economic benefits produced and transferred out of those counties. [9–3] Their views were received by some people as a personal attack on the West and their livelihood.

Sarah Vogel, the elected Commissioner of Agriculture for the state of North Dakota, was among leaders in the Northern Plains who saw the book in a different light. "You've heard of **Buffalo Commons**?" she asks. "Our response was the Bison Cooperative. We had 180 ranchers pool together and start their bison business. They now sell bison meat to fancy restaurants on the East Coast for a lot better prices than they get for their cattle!" [9–4]

Commissioner Vogel laughs at the paradox of building a bison raising and processing cooperative in the heart of **Buffalo Commons**. It is poetic. And it is a perfect example of the new generation cooperatives being formed on the Northern Prairies that can rightfully be called "fourth wave" development, as discussed in the previous chapter.

Buffalo Commons served as a call to action in North Dakota. The Decline of Rural Minnesota, a book of essays looking at Census data and trends in western Minnesota counties by Southwest State University historian Joseph Amato and Canby, Minnesota, city administrator John W. Meyer, served the same purpose for communities in what could become "corn and soybeans commons." [9–5] At Renville, in southwestern Minnesota, farmers responded by creating new generation cooperatives that have touched off developments similar to a chain of events.

These positive developments wouldn't need to happen. Southern Minnesota gets about four times as much annual rainfall as the central counties of the two Dakotas. Like Iowa counties to the south, it contains some of the nation's best corn and soybean land. But as lush as it can be, the land east of **Buffalo Commons** is no more hospitable for today's rural population. Left to evolve under the economies of modern crop farming, there is no reason why contiguous corn and soybean fields won't stretch all the way from La Crosse, Wisconsin, to

Sioux Falls, South Dakota, interrupted only by the Mayo Clinic at Rochester, Minnesota.

Go south a few miles and the same could be said for the expanse of Iowa reaching from the Quad Cities on the Illinois side of the Mississippi to Omaha, Nebraska, on the west; or Missouri, from East St. Louis, in Illinois, to Topeka or beyond in Kansas.

The Amato–Meyer book has helped prevent such a fate in southern Minnesota, says Jim Norman, Renville city administrator and member of Renville County's Economic Development Commission. "We bought a few of those books. We passed them around. They've been in a lot of different hands," he says. [9–6]

***

The Amato–Meyer book served as ammunition for people in and around Renville who were proposing new ventures and arguing for change. Area farmers and townspeople at Renville and neighboring towns of Olivia, Danube and Sacred Heart had actually started the process of rebuilding their area economy and larger, shared community a few years earlier. They invested $31 million in the early 1990s to launch new cooperatives and joint venture businesses to boost the value of area crops. [9–7]

Their investments were part of an explosion of more than $1.1 billion in similar capital expenditures farmers and rural townspeople were plowing into new ventures in a band reaching from Wisconsin through the Dakotas, mostly within the period of 1992 through 1994. [See Chapter Note A] Dennis Johnson, president of the St. Paul Bank for Cooperatives, said preliminary planning suggested the amount of new investment could double by the end of the decade. But he also noted that most such new generation cooperative development, as defined in the previous chapter, was still isolated to the Upper Midwest states that could be called "New Scandinavia." Leading the charge in building new ventures were the grandchildren and great–grandchildren of Northern European immigrants – Scandinavians and those from Germany, Holland, France, Belgium and the British Isles – who brought ideas about cooperative enterprise to the Northern plains. And they were also the children of farmers who earlier gave rise to such regional co–ops as Harvest States, Land O'Lakes and Cenex.

The new co–op ventures ranged in size from large to small. The largest under development is a $245 million corn sweetener processing joint venture – ProGold Co. – planned for the Red River Valley area around Fargo, North Dakota and its twin city, Moorhead, Minnesota. It was formed by two existing sugar cooperatives and a newly formed corn growers co–op, Golden Growers. Lower cost ventures included a new marketing co–op for organically–grown vegetables near the Lake Superior resort communities of Ashland and Bayfield, Wisconsin, and a farm–raised venison cooperative at Morris, Minnesota. They needed less than $100,000 in member equity for startup costs and no initial plant investment.

The new businesses were varied, too. Some co–op ventures feed farmer–owners' crops to pigs, cattle and chickens in large–scale operations that then sell meat or eggs. Others use crops to produce ethanol fuel in local plants. Some cooperatives invest in huge plants and the latest scientific and technological developments in an effort to be the lowest–cost producer and do large–volume business. At the other end of the spectrum, many co–ops intend to stay small and serve niche markets for such products as milk from cows not treated with growth hormones.

"Farmers are getting up off the ground," proclaims Dana Persson, president and general manager of Co–Op Country Farmers Elevators that is headquartered at Renville. "They are moving farther up the food chain so they aren't totally dependent on commodity prices for their income." Adds Wayne Brandt, executive vice president of the Minnesota Forest Industries trade association at Duluth, "Agriculture is doing in the 1990s what our industry did in the 1980s. While many of the same forestry companies were fighting spotted owl habitat issues and environmental restrictions in the Pacific Northwest, they invested $2.2 billion on new plants and technology in Minnesota and more than $1 billion on similar improvements in Wisconsin to leverage the value of raw logs. Similarly, he notes, "the farmers are now trying to raise the value of their natural resources at home, in their own communities."

Renville area farmers have been at the forefront of these new developments. Three decades ago they started Southern Minnesota Beet Sugar Cooperative, which is just outside of town. A decade ago, they used the sugar co–op's model to help start Minnesota Corn Processors Inc. at Marshall, Minnesota. Their new round of co–op building was started with the merger of four area grain elevator companies into Co–Op Country in the late 1980s. That merger brought efficiencies to grain handling, and the new co–op began looking for ways to diversify and raise the value of area crops. But its members voted down its first diversification plan, which would have started a large swine breeding and feeding program to use area corn.

"We got clobbered," Persson recalls. Members said they didn't want their large area grain co–op to be in direct competition with their own livestock production. Their argument gained support from members of the Minnesota Farmers Union and some elements within the Minnesota Farm Bureau, and from some prominent representatives of the dominant Catholic and Lutheran rural social service agencies.

But members of Co–Op Country who saw new generation co–ops as a way to keep people on the land, not drive them off, used the parent co–op's swine breeding model to start ValAdCo in 1992. It was expanding its operations in 1994 to house 8,750 breeding sows. When completed, the ValAdCo farmers will have $17 million invested in modern barns and equipment.

And east of Renville, at Olivia, a second set of farmers pooled resources to start Churchill Cooperative, another swine co–op known as a "multiplier unit" that was patterned after the ValAdCo model.

With the large–scale swine breeding and feeding ventures launched, area grain farmers from the Co–Op Country membership formed Midwest Investors to start an egg industry at Renville. Butch Buschette, vice–chair of both the Midwest and Co–Op Country boards, says the egg co–op's purpose, like that of the swine ventures, was to add value to members' crops. "We would rather truck eggs out of here than haul out the feed you need to produce eggs," he says. But he sees great potential to move Midwest's farmers even higher on the food industry ladder.

Midwest was building barns valued at $9 million on the east edge of Renville during the summer of 1994. Within a year it intended to have one million laying hens producing eggs. The co–op will market the eggs through a joint venture business arrangement with Primegg, a private egg company at Madison, Wisconsin. The initial stage was expected to produce annual egg sales of $100 million, with profits of $4 million to be distributed to its investors. But Buschette says plans are already in place to keep building barns and expand to two million birds, and the co–op may build a plant to further process eggs into products for use by food industry companies – moving Midwest Investors even farther up the ladder towards the consumers' grocery carts.

ValAdCo and Midwest greatly expanded the Renville area livestock industry. That, in turn, expanded the local market for sophisticated feed mixes. So ValAdCo and Midwest joined with Mr. Persson's Co–op Country to form another joint venture, United Mills, to supply hog and chicken feed. United Mills opened a new $1.8 million feed mill business in Renville in early 1994. Co–op Country, meanwhile, expanded its services in step with its members' new ventures. It was providing accounting and administrative services for the new co–ops. And Mr. Persson, who had visited the Netherlands in 1992 to study its co–ops and farm structure, borrowed a forward–thinking farm management and environmental idea from the Dutch "manure bank." (See Chapter 4) He started a manure management service for the new co–ops, making manure a part of the fertilizer mix Co–Op Country sells to area farms.

Despite its enormous size, Midwest's applications for permits to build chicken barns cleared Minnesota Pollution Control Agency review in record time. "They (the agency) saw we had all these thousands of acres of land for manure disposal and a plan to manage it," says Mr. Persson. "It had to be the best plan they'd looked at."

All these additional services means Co–Op Country is growing fast. As a grain elevator co–op, Co–Op Country and its predecessor associations continue to be federated members of Harvest States Cooperatives. It became an affiliate of Cenex as it expanded into fertilizers, farm inputs and agronomy services. And its United Mills joint venture, meanwhile, has aligned itself with Land O' Lakes for feed nutrition services. Meanwhile, Co–Op Country outgrew its suite of offices at the Renville elevator. It built a $300,000 headquarters building in Renville, also in 1994, to house its accountants, managers and field representatives and planned to lease its smaller elevator offices to Midwest Investors.

Construction crews were busy at farms around the Renville area as well. Farmers began modernizing their own livestock and poultry operations with private financing. Some farmers expanded and built modern facilities to supply the new co–ops. South of Danube, for instance, Tony and Barb Frank built new poultry barns valued at $1.5 million to raise laying hens for Midwest. The Franks secured a 10–year contract to supply pullets for Midwest's egg barns, taking day–old chicks and raising them for 18 weeks until they became layers.

"We've raised pullets before, but we couldn't expand. We've trucked pullets as far as Missouri and Washington state because we didn't have a local egg industry," explains Tony Frank. But that all changed with Midwest's arrival, adds Barb, his wife and partner. "You can take the supply contract down to the bank – it's money. Without the co–op, I'm not sure we'd be here."

Current agricultural and rural demographic trends strongly suggest the Franks would have moved off the land without Midwest's need for pullets. They are in their 30s, as are most of the new cooperators in Midwest Investors, says Mr. Persson. ValAdCo's initial investors were only a few years older, with an average age in the early 40s. While these younger farm operators were anchoring themselves to the land with new generation co–ops, Minnesota kept at its pace of losing about 2,000 farm units a year. In mid–1994, the Minnesota Agricultural Statistics Service noted that the number of farms in the state had fallen to 85,000 at the start of the year. That was down from 104,000 farms in the early 1980s. Other farmers, whose crops are raised for export or shipment to feeders and processors in other communities and states, keep expanding by taking over their neighbor's land.

The Renville area's new generation cooperatives were helping members expand up the food chain and provided them economies of scale in production. By doing so, the co–ops reduced the need for their members to expand horizontally by building massive individual land holdings.

ഇ ഇ ഇ

Investment in these new ventures varies with the co–op and its type of business, explains Mr. Persson, with farmers paying from $5,000 to $15,000 for a share in the large–scale businesses. Each share would grant the member an equity right to sell a certain amount of a field crop to the co–op and derive a proportionate share of the value added by the company's business. The farmers who formed Midwest Investors, for instance, bought shares valued at $7,000 each. "Farmers usually translate their investments into bushels of grain, such as corn here and wheat at the Drayton (North Dakota) pasta co–op," Mr. Persson explained. But "dollars or bushels," he says, the investments can be large and often require farm financing from local banks or AgriBank, the St. Paul–based cooperative farm lending bank. With the farmers' initial investments for startup money and leverage, the new co–ops then can turn to big banks or the Farm Credit System's banks for cooperatives to finance construction of large barns and plants.

Then the new co–op venture begins adding value to the farmer–owner's crops. At ValAdCo, for example, the corn the co–op owners grow was worth slightly less than $2 a bushel in mid–1994 when bumper crops were waiting in the fields. A hog raised in modern swine facilities would eat less than 10 bushels of corn before reaching market weight. So – before factoring in costs of buildings, labor and management – the corn farmers raise the value of about $20 worth of corn to about $95 by producing a market–ready hog. That same hog would draw premium prices from meat packers who want both the uniformity of hog size and the volume of animals such an enterprise can provide.

Of course, the impact is less dramatic after production costs are accounted for, Mr. Persson noted. But he added that while the farm prices for corn and hogs weren't good in 1994, ValAdCo's farmers were making money. Iowa State University and the U.S. Department of Agriculture estimated at the time that average costs of raising hogs on

223

Midwestern farms – from Ohio to the Dakotas and Nebraska – was $44 to $45 for 100 pounds. Large–scale, modern operations had production costs of around $37 to $38 for 100 pounds. Market prices for hogs – before premiums were paid to the large operators – were in the low $40s for 100 pounds. Multiply the numbers out to market hogs of from 220 to 240 pounds and the new co–op hog farmers were making from $4 to $10 per animal while most traditional hog farmers in the Midwest were losing money.

෩෩෩

By itself, each venture at Renville represents classic economic development. They create jobs – about 100 new jobs in total – and they spur other economic activity. Earnings, or profits, generated by the value–added activity at Renville, stay in the community and aren't transferred out to corporate headquarters elsewhere in the United States or abroad. This additional income "helps everybody on Main Street and everyone in the community," says Allen I. Olson, a former North Dakota governor who now heads the Independent Community Bankers of Minnesota. "This is where economic development crosses over and becomes real community development," he says. [9–8]

Gary DeCramer, fellow of the State and Local Policy Program at the Hubert H. Humphrey Institute of Public Affairs, goes even farther:

"If the first three waves of development strategies sought economic development, this new wave seeks community development." [9–9]

This writer and observer of rural development would argue that cooperative development that looks vertically as well as horizontally, whether producer–owned, worker–owned, or a joint venture bringing together various community interests, is the "fourth wave". It meets the criteria suggested by the Chicago Federal Reserve Bank economists, as outlined in Chapter 8; it copies the complete ground–up development of food systems that are being created in Third World countries and former Soviet states, as described in Chapter 6; it adjusts American agriculture to changes in the world food system as described in Chapter 2; and it positions local communities to both partner and compete with the sophisticated European co–ops and multinational food companies described in Chapter 4. Most importantly, the new generation ventures equip farmers and their townspeople neighbors to give communities a new lease on life – a fighting chance to survive into the Twenty–First Century.

For Richard Magnuson, the veteran Doherty Rumble and Butler cooperative law expert who has helped give shape to many of the new ventures, the changes occurring in rural America are like a new planting season. "It's as if everyone is out planting, and everything is turning green," he says. "It's springtime again."

ഇഇഇ

Jim Norman, the Renville city administrator, sees the dawning of a new era for his town. The farmers' new ventures are stimulating community development. Few people in western Minnesota would have predicted it a few years ago.

Renville is a farm service town of about 1,300 people wedged among larger farm service and rural manufacturing towns of Olivia, Willmar, Granite Falls and Redwood Falls. The pattern for towns such as Renville, as noted in the Amato–Meyer book, is the gradual loss of farm service and retail functions to the bigger nearby centers. Smaller farm service towns graduate into the role of senior housing communities for retired area farmers and townspeople. Since rural population trends don't support the perpetuation of such communities, there are questions about how many more decades these communities can survive.

But Renville is growing again. The city has bought a farm on the east end of town and made it a large city industrial park. Midwest Investors' barns are the first occupants in the park. The city also is developing land on which 25 to 30 single–family homes will be built – a project not even dreamed of 10 years ago.

"Let's just say that you aren't unemployed out here if you can string wire, lay pipe or operate a hammer," says Administrator Norman.

The town of Carrington, North Dakota, between Fargo and Jamestown, is riding a similar boom. Three new generation co–ops have started there since 1990, and Carrington has a severe housing shortage that is spurring home construction, says Jack Piela, of the North Dakota Coordinating Council on Cooperatives. Area wheat farmers started the Dakota Growers Pasta Co–op, which makes pasta products from their durum wheat in a $41 million plant. Other farmers built a $640,000 fish–farming business, Dakota Aquaculture Cooperative. A third group has formed Northern Plains Organic Co–op to build a $1.5 million flour mill to make niche market flours from their organically grown grains.

By mid–1994, about 20 different new generation cooperative projects were being studied for possible building within the next few years, said the Council's Piela. "If they all go forward and succeed, we will probably have $1 billion or more in new development just in

North Dakota," he said. But that's a big "if" – in North Dakota or any rural, agrarian state.

∽∽∽

Al Oukrop, who operates a cooperative business consulting agency at Burnsville, Minnesota, says he doesn't know how many new cooperatives hire independent feasibility studies and do professional market research before launching new ventures. This concern has been echoed by others involved with cooperative businesses. The U.S. Department of Agriculture was being reorganized in the mid–1990s in ways that should help provide coordination of rural development services, notes former Agriculture Secretary Bob Bergland. At a minimum, he says, the reorganization should help various USDA agencies work more effectively as a partner in rural development projects. The extent to which these agencies provide assistance to new co–op ventures, of course, will be determined by Congress in future farm bills.

That notwithstanding, there are a variety of public, private and cooperative support agencies in place to help new ventures. Nationwide, Cooperative Development Services does provide market studies and financial planning services for new co–ops and a CDS Subsidy Fund, established with grants from established cooperatives, helps prospective ventures secure technical assistance. Similar support to local entrepreneurs and enterprising cooperatives comes from large regional foundations, such as the privately founded McKnight, Bremer and Northwest Area foundations.

Some land grant universities have cooperative centers and research capabilities that connect with local co–ops through the federal–state Extension Service. Jim Arts heads a multi–state Cooperative Development Services office at Madison, Wisconsin, that was working with experts from Iowa Extension Service at Iowa State University in the mid–1990s to support creative projects being planned at several Iowa communities. Washington, Oregon and Minnesota state governments have public–private research agencies to help entrepreneurs and co–ops develop and utilize local resources. Washington even has a state office to help workers buy plants through employee stock option plans – taking the concept of new generation cooperation to an important new level.

North Dakota's special state agency, the Agricultural Products Utilization Commission, has been expanded in recent sessions of the North Dakota Legislature to make it more useful for starting new generation cooperatives, says Sarah Vogel, the state agriculture commissioner who serves on the commission. That group makes grants to new projects for legal fees and feasibility studies. It also provides grants for initial organization expenses. Eventually, large banks and cooperative lending institutions that might finance construction of plants also serve as a backstop for these projects, Commissioner Vogel says. Lenders want to see both marketing and business plans before helping local groups start new ventures.

But no central agency, trade association or research group is monitoring all new co–op developments – in part because the cooperative movement in the United States is fragmented among rural and urban co–ops, producer and consumer co–ops, new generation service co–ops, such as health care and senior housing projects, and community service co–op projects, such as starting and building golf courses to improve the quality of life in rural communities.

Gnawing at some co–op watchers' stomachs in the mid–1990s are unanswered questions about the potential size of markets for some new generation co–ops. What happens when farmers outside the Upper Midwest aggressively start building similar new enterprises? Corn farmers have several ethanol co–ops on their drawing boards. [See Chapter Note A] There is a federal ethanol program only because Vice President Albert Gore supported the farmers in a battle with the oil companies and broke a Senate tie vote in 1994. How will competing agricultural processors such as Archer Daniels Midland and Cargill respond to these same markets the farmers are targeting?

Allen Gerber and Frank Blackburn at the Minnesota Association of Cooperatives have gathered extensive data about the emerging new generation co–ops. But they and William Nelson of The Cooperative Foundation see an urgent need for a new depot, or terminal, along the co–ops' new information highway. There, farmers, community leaders and other planners of new cooperatives could access information on available research, past models, and the opportunities and pitfalls that will challenge their plans. But the birth of new generation co–ops is so recent, and so regionalized, that nothing is in place.

කකක

Joe Famalette, president of American Crystal at Moorhead, Minnesota, says he feels pressure to make sure the ProGold corn sweetener venture gets off to a successful start. It will be watched closely and will serve as a model for other farmers, just as American Crystal was a model 20 years early, he said. American Crystal, Minn–Dak Growers Cooperative at Wahpeton, North Dakota, and Golden Growers, a new corn growers cooperative involving farmers from

Minnesota, North Dakota and South Dakota, were planning to select a site for their new corn plant by early 1995. It would be built in one of the three states near the Fargo–Moorhead area. And its corn fructose products would be marketed through United Sugars Co., a new, joint venture at Edina, Minnesota, that was formed to sell sugars from the two Red River Valley co–ops and Southern Minnesota Beet Sugar at Renville.

"Any success we have will be copied by other farmers," Mr. Famalette says. "I think the burden of doing this corn thing right is bigger than the project itself."

There can be no doubt that he is right. American Crystal's co–op was born by farmers pooling resources to buy sugar beet factories a private company was planning to close. Had the plants closed, they would have lost their market for beets. By buying the plants and starting their own company, they preserved their market and began earning income from sugar processing. This model is inspiring other farmers in the Upper Midwest. Since United Sugars is branching out to market corn sweeteners as well as sugars, farmers are turning to Mr. Famalette to discuss possible other ventures. He's been approached by farmers planning to start co–ops for raising and processing carrots, dry edible beans, potatoes and specialty wheats who want to "bundle" their products with United Sugars' sales force. "I think we are seeing the evolution of a new agriculture, of farmers taking control of their futures," he said.

It was far too early at the time of this writing to predict if anything would come from those talks. But the American Crystal board of directors and management is clearly in the market for expansion and diversification. A delegation from the co–op visited the Dutch cooperatives in July, 1994. After returning, Mr. Famalette said he and former North Dakota Gov. George Sinner, who is also an American Crystal executive, spent a weekend together discussing what, in effect, are the lost opportunities of American cooperatives.

"George (Sinner) remembers from when he was governor that 60 percent of U.S. barley is grown in North Dakota. A barley mill has expanded four times out here. But the price of barley for the producer is the same now as it was 30 years ago," Mr. Famalette said. "You don't expand capital investment four times unless you're making money. We should have owned the barley milling!" [9–10]

৩৩৩

There is an explosion of new cooperative ventures in the Upper Midwest states, or "New Scandinavia" as St. Paul Bank for

Cooperatives President Dennis Johnson described it. It does draw from past successful models and from the cultural and ethnic heritage of the Northern European immigrants who settled in the region. But these new ventures draw from other American experiences and from currently changing world markets as well, as noted by other leading agricultural cooperative experts:

ഇരുഇ

Mike Cook, the agricultural economist who teaches cooperative management courses at the University of Missouri at Columbia, says new generation co–op building was almost exclusive to the Upper Midwest in the mid–1990s. When speaking to farm leaders in the Missouri, Kansas and Arkansas area, for instance, "I keep telling people to keep an eye on what's going on up there."

At the same time, he says, the new generation cooperatives of the 1990s are closely akin to the specialty crops co–ops formed earlier in California. The Blue Diamond Growers, for instance, and Sunkist, are examples of growers starting co–ops to process and market their own produce in an "offensive" marketing strategy. Ocean Spray cranberry growers in Massachusetts and Wisconsin are another good example.

Most large regional cooperatives were formed for "defensive" reasons such as countering perceived problems with marketing or buying farm supplies. [9–11] Professor Cook sees five primary motivations for starting co–ops, of any era or different business structure:

"These include: (a) the desire to avoid the negative consequences of market power; (b) the drive to attain scale economies in procurement, services or marketing; (c) the attempt to reduce risk; (d) the quest to provide missing services; and (e) the drive to achieve additional margins. The first four of these are drawn from the market failure literature and the fifth from the private approach to managing excess supply created price levels." [9–12]

Over time, most of America's large regional cooperatives became hybridized with operations addressing all five goals cited by Professor Cook. Farmland Industries' meat operations, Harvest States' milling and Holsum Foods units, and the dairy foods units at Land O'Lakes, Associated Milk Producers Inc., Dairymen Inc. and Mid–America Dairymen in the central states, and Seattle's Darigold Farms cooperative are examples of value–added enterprises that seek returns for dairy farmers that are greater than the raw value of their milk production. [9–13]

But Professor Cook has also noted that cooperatives lag behind private and publicly–traded food and agribusiness companies that are pursuing growth in the global market. While American exports of bulk farm commodities boomed in the 1970s, the export value of processed food and feeds and consumer food products moved ahead of commodity trade in the 1980s. American co–ops are just starting overseas operations and aligning themselves with multinational partners to move their products into global markets in the 1990s. [9–14]

ಜಜಜ

Mike Boehlje, agricultural economist at Purdue University in Indiana, says he also is telling farmers in the Eastern Corn Belt to watch new generation cooperative developments in the Upper Midwest states. "It's pretty obvious when you look at changes in the hog industry, for instance, that there are two models for the future," he says. "Either you have farmers banding together to use the latest genetics and technology, like in parts of Minnesota and Iowa, or you have food companies doing it themselves, like we're seeing happening in North Carolina and Southern states."

In mid–1994, the U.S. Department of Agriculture supported that view in its **Situation and Outlook** report. [9–15] It forecast commercial pork production at 17.37 billion pounds for 1994, a new record, and forecast that 100 million head of hogs would be slaughtered in 1995, yet another record. But it also noted that herd expansions to produce these record numbers were occurring in North Carolina and Missouri, where vertically integrated food companies raise the hogs or have large–scale farmers raise the animals for the companies under supply contracts. Hog production in the traditional Midwestern pork producing states was actually in decline.

"Herd cutbacks among smaller producers are expected to continue," the report said. "Thus, higher–cost producers are expected to experience a cost–price squeeze over the next year, which may force many to exit the industry." [9–16]

It is part of the trends reshaping American agriculture, says Purdue University's Professor Boehlje. He has become one of America's most popular speakers on agricultural economics and development issues by defining "megatrends" changing agriculture by the year 2000. The sum of those trends, he adds, is that "family farmers of the future will be defined by their management of labor, production and marketing opportunities and less by their own labor and land holdings." [9–17]

Clearly, new generation cooperatives that move farmers farther up the food chain are one cost–effective way for producers to accommodate the trends in agriculture. But for his own state, Professor Boehlje doesn't see production co–ops playing a large role in tomorrow's Indiana. In 1992, he and his colleagues at Purdue published a comprehensive study, **Indiana Agriculture 2000**. Among predictions for the structure of agriculture in their state, they said: "In spite of the potential role they could play in coordination of the food production and distribution system, cooperatives will likely not play a much larger role relative to non–cooperative firms in the agriculture of the future." [9–18]

ಐಐಐ

Robert Cropp, the agricultural economist at the University of Wisconsin at Madison and director of the Wisconsin Center for Cooperatives, has told members of the National Farmers Organization they have a big decision to make during the decade of the 1990s. These independent farmers must decide who they want to work for if they choose to farm in the Twenty–First Century. Management of production is shifting, he said at the NFO's 1994 annual meeting at Ames, Iowa. Either they will work together to farm for themselves, or they will be farming for someone else – most likely an integrated food and agribusiness company.

It was a dramatic way to speak to a farm group. But later, while attending a conference for cooperative educators in Minnesota, Professor Cropp said his NFO speech was a call for farmers to think about how they will accommodate the changes occurring to the structure of agriculture. The NFO is best known on the periphery of agriculture as a broad–based farm organization. Structurally, however, it is a bargaining association that pools members' livestock and crops for volume marketing similar to a cooperative. Its public policy pronouncements attract wider attention from the general public than the work it does for its members. Nevertheless, its marketing services add value when it brings higher prices to its members through forward contracting sales to packing plants and processors, and when it is rewarded for volume. In spite of such benefits, it hasn't arrested the decline of small, family–run farm enterprises in its service territory. Nor could any other farm organization or single co–op.

Professor Cropp and his academic colleagues at Madison have issued similar warnings to other farmers about their need to adjust to change. They forecast to Wisconsin farmers and state officials that "America's Dairyland" would lose one–third of its dairy farmers in the 1990s, just as it did in the 1980s. Moreover, they warned, the decline in dairy farm numbers would not come from horizontal expansion by large farmers, as is happening in grain country. Rather, the small dairy farm is being lost to attrition; young farmers do not take over and perpetuate dairy farms designed for 40 to 60 cows when older farmers retire. So into the mid–1990s, Wisconsin was witnessing the loss of dairy farms and the decline in dairy production; California moved ahead of Wisconsin in milk production in 1993.

"We're starting to see young farmers getting together and looking at large, 600–head cooperative dairy projects. Some larger," he says. He worked with a group of young farmers in the early 1990s and helped prepare studies for them. They backed down at the last minute. But he's working with other groups that are looking at the economies and efficiencies of sharing large–scale, modern facilities similar to those used by large private dairy companies in California, Arizona, Texas and Florida.

The states of Wisconsin and Minnesota both launched study commissions in the 1990s that were instructed to find ways to resuscitate their respective dairy industries. New state financing programs were being devised to help dairy farmers invest in new dairy ventures in both states. Dairy production co–ops are coming, Professor Cropp predicts, if for no other reason than the lifestyle burden of dairy farming. Participating families in a new generation dairy co–op can schedule vacations and have some weekends free of dairy chores each month. Or, as he once explained to a group of Midwest co–op directors, future cooperating farmers "aren't married to their cows."

One reason why dairy production ventures may be lagging those for pork, chicken and beef might be the cooperative history of the dairy industry. "Most of the dairy farmers are already members of co–ops. That may by holding them back from starting new co–ops to produce the milk they are already selling through co–ops." [9–19]

ଚଗଚଗଚଗ

Harlan Hughes, Extension livestock economist at North Dakota State University, has written in newspaper columns published across both Dakotas that the beef, pork and poultry industries are all undergoing similar changes, and for the same reasons. "The primary force driving these changes is the need to decrease per–unit costs of meat production," he says, noting that reducing production costs is key to gaining market shares, and the animal industries are waging battles with consumers to gain larger market shares." [9–20]

And in an earlier article for newspapers in the Dakotas, Professor Hughes concisely explained why farmers shouldn't argue about whether they should change their farming practices. Rather, he explained why they must. A portion of his column is excerpted here:

"The structure of animal agriculture is changing. The number of beef cow herds is decreasing while the remaining herds are getting larger. The dairy industry is changing from the small one–family herds toward multiple family herds. California, with its large dairy farms, has now become the number one dairy producing state in the nation. These major structural changes in animal agriculture are the direct result of the increased economic efficiency of the larger, specialized production units.

"Over the past 15 to 20 years, farmer cattle feeders have shut down their small Corn Belt feedlots in large numbers while the Central and Southern Great Plains have been adding large cattle feeding lots. The center of the swine industry is shifting out of the Corn Belt to North Carolina and other Southern states which welcome larger farms. Structural changes in agriculture are absolutely going on all around us.

"Most of today's structural discussions tend to focus on the swine industry. Sources indicate that large swine operations such as Carroll Foods and Murphy Farms control approximately 15 percent of the U.S. swine production and they are expanding 15 to 25 percent per year. In a few years they will control 25 percent of the swine production – enough to fuel a profound change in the traditional hog cycle. While Iowa is still the largest swine–producing state with 1.65 million sows, Iowa's percentage of the national swine herd is decreasing.

"The food processing industry is increasing its focus on the production of highly consistent food products geared toward consumer demand. Consumers are demanding consistent, wholesome, nutritious and safe food. All of this puts an economic premium on integrated food production systems that control food quality and consistency from the farm to the consumer. ...

"The primary implication for production agriculture of this new focus on food product quality and uniformity is the loss of farmer and rancher independence. In this new era, economic rewards will increasingly go to those commercial farmers and ranchers who produce specialized, uniform agricultural products sold under contract specifications to a food processor. ..." [9–21]

ಬಂಬಂಬಂ

At the University of Saskatchewan and its Centre for the Study of Cooperatives, historian **Brett Fairbairn** is keeping watch on cooperative developments across the border in the old Northwest states. Though the Western provinces developed along similar historical paths as their neighboring states, co–ops in Saskatchewan are taking different paths to future development. Canadians appear to be ahead of their American cousins in developing health care and housing co–ops that are emerging to address quality of life and community service concerns in both countries. And Canadians also seem to be farther ahead in forming worker co–ops to take over and operate plants.

But they lag behind the Upper Midwest states in forming new generation cooperatives aimed at value–added processing or cooperative production. [9–22] Professor Fairbairn, however, says that may quickly change. Groups of Saskatchewan cooperative leaders took bus tours of North Dakota co–ops in 1994 to study new ventures being formed by the American neighbors. [9–23]

In a profile of Saskatchewan cooperatives, the Centre shows the province's residents have used co–op business structures for many of their community services. The province has 1,400 cooperatives spread over a wide range of business and service ventures. Of them, 25 percent are agricultural and resource co–ops, 25 percent are financial institutions such as credit unions, 19 percent are recreational enterprises, 13 percent are retail and wholesale businesses, and nine percent are child care and educational co–ops. [9–24] The remaining nine percent of the province's co–ops are divided among community service ventures, such as health care clinics, and community development co–ops formed by rural communities, farmers and townspeople.

"The formation of new cooperatives and the maintenance of existing ones can provide rural areas with an opportunity to sustain and revitalize their local economies," the Centre's report concludes. "In an era of relatively high unemployment and limited economic opportunities, cooperatives provide Saskatchewan with an important avenue for community and economic development." [9–25]

ৰেৰেৰে

Revitalization and community development, as described in the Canadian study, is starting to spread through the Midwest and Western U.S. states. New generation cooperatives are changing rural community fortunes. And so are older, existing co–ops that are using creative tools to reach out and help revitalize their communities. These companies and communities are restructuring old co–op models to serve new purposes.

Lani Jordan, spokeswoman for Cenex, points to a Cenex affiliated petroleum co–op at Black River Falls, Wisconsin, as an inspiration for other local co–ops. Seeing both community needs and opportunities, the co–op began expanding in the mid–1980s. It now has a restaurant that serves both the community and travelers along Interstate 94. That led to operating a Best Western motel. Then, when a nearby strip mall was in financial trouble, the cooperative took it over and now leases space to an art gallery and other retailers who serve interstate traffic as well as enrich the local community.

South Dakota writer David Aeilts tells how farmer–members of James River Farmers Elevator at Groton pitched in with the South Dakota community's townspeople to build a $500,000, nine–hole golf course on the edge of Groton. The golf course venture helped strengthen the broader community, bringing town and country together in collaborative action, he says. [9–26] And while this venture was taken for quality of life purposes, the golf course is credited with raising Groton property values by 25 percent. [9–27]

Within three years, the Groton experience inspired others. People in the Pine Island area on the south edge of the Minneapolis–St. Paul metropolitan area were busy raising money and pledging sweat equity in building a cooperative, community golf course.

అఅఅ

The large regional cooperatives have become well known to consumers for their brands of products and logos painted on community buildings and gas pumps. What isn't as visible is the help co–ops such as Farmland, Cenex, Agway, Land O'Lakes and Harvest States give local affiliate associations to cope with change. Defensively, this help might be legal and structural assistance as two or more local grain elevator associations merge into a more efficient unit. Offensively, it is help in finding ways for a local co–op to expand, both vertically or horizontally, to improve its return on equity for its members.

As an example, Lincoln Mutual #2 at Reardan, Washington, has come back from financial hard times in the 1980s. A big part of its recovery is credited to reinvigorating the farm supply co–op's own membership. [9–28] Education and training programs to help members understand their co–op, and grasp that they own it, are part of the process. Reaching out to young people and new residents is another. "It's hard to think of one young farmer in that area who hasn't graduated from Cenex/Land O'Lakes cooperative education programs," says Bob Holloway, Cenex/Land O'Lakes training specialist, who led those programs for 17 years." [9–29]

Along a similar vein, the Cenex/Land O'Lakes publication has shown the co–ops' respective directors how three of their affiliated locals strengthened balance sheets by starting or expanding new, niche market services to their communities. Cenex Supply and Marketing, at Tangent, Oregon, bought the equipment and customer list from a local company for $105,000 to launch a residential and commercial lawn care service. With the co–op, the customer list has nearly doubled and annual sales have grown to about $170,000. In North Dakota, Farmers Union Elevator at New Salem started Rough Rider Feed and Supply at nearby Mandan to supply specialty feeds. The new business supplies about 10 tons of rabbit feed a month to area rabbit raisers, and it is building a wild bird feed business by hosting bird feeding clinics for area urbanites. And at Cadott, Wisconsin, and Juniata, Nebraska, local petroleum co–ops associated with Cenex have built diversified truck stops along nearby major highways to expand their businesses and local business services.

The River Country Co–op at Chippewa Falls, Wisconsin, for instance, built River Country Plaza that has parking and services for up to 70 trucking rigs, provides LP and dumping services for tourists' RVs, and an 140–seat restaurant that has banquet facilities to serve area groups of up to 80 people. The co–op used technical and marketing advice from Cenex, commissioned three feasibility studies, and hired an auditor and external consultant before launching its $2.9 million project. [9–30]

ಇದ್ದ

Finally, there are other models around to guide rural community leaders and farmers who may explore ways to revive and reshape their area economy for the Twenty–First Century. Some of these models come from another important frontier as we approach a new century – America's big cities – involving credit unions, housing cooperatives, worker co–ops and community development cooperatives.

Other models come from Southern Europe, not Northern Europe. And still others come from new creations, the partnering and joint venture arrangements now being shaped worldwide in a global economy that link cooperatives with other business entities. Regardless of structure, all of these models depend on a cooperative spirit in which the partners work for a mutual good.

ಇದ್ದ

People involved with rural development in Iowa and Washington state are exploring different avenues to development than the paths now being taken in Minnesota, Wisconsin and the Dakotas. **Roger**

Ginder, agricultural economist at Iowa State University, gives an explanation why this is so. First, he notes, Iowa farmers don't have the problem of being far from markets. Archer Daniels Midland has a large processing complex at Cedar Rapids and Cargill has large processing operations at Eddyville. Cargill is also developing a large complex at Blair, Nebraska, which will also provide a convenient, nearby market for Iowa farm commodities. "It puts us (Iowa) in a little different position for capacity. The farmers who started Minnesota Corn Processors were farther removed from markets," Professor Ginder says.

And that is an important difference. Farmers pay the cost of shipping grains and oilseeds to markets in what the grain trade calls "basis points." Usually, it means that farmers in Minnesota, the Dakotas, Montana and Idaho are paid less per bushel than the market price listed at commodity exchanges, ports and cash grain markets. The difference compensates for moving the grain to where it can be used or exported. Rarely, but occasionally, the basis will be "positive" as the grain trade pays premiums to pry grain out of storage and onto the market.

Regardless, Professor Ginder sees Iowa farmers' closeness to processing plants and nearby Mississippi River barge terminals as a reason why farmers haven't explored alternative ways to create markets. [9–31]

But it is starting to happen in Iowa, too, he says. Jim Arts of Cooperative Development Services, as mentioned earlier, was meeting with farmers in the mid–1990s who may pool resources to build large–scale swine multiplier units like those operated by swine production and marketing companies in North Carolina and by farmers at Renville, Minnesota. And on a different path, Iowa farmers have pioneered in working a partnership arrangement with a private company – in this case the world's largest private agribusiness company.

Agri Industries, a grain cooperative based at West Des Moines, formed a partnership with Cargill Inc. in 1986 when the co–op was experiencing hard times. The partnership arrangement was reviewed and extended until the year 2006 in a 1994 vote by the farmer–members. [9–32] The partnership, which operates Agri Grain Marketing (AGM), is a classic example of the partnering described by ADM Chairman Andreas in Chapter 2. "This partnership is an outstanding example of how a cooperative and a private business can work together successfully," said Bob Neal, general manager of AGM. "Agri's strengths (grain origination) and Cargill's capabilities (marketing) have yielded a unique and mutually beneficial combination." [9–33]

Iowa is also exploring ways to combine its local manufacturing base with area agricultural strengths. New generation co–ops and partnership arrangements, such as described above, will be important components in forming new community economic arrangements. Most likely, local and state governments will also become full–fledged partners beyond simply offering tax incentives.

A good model for these new arrangements comes from the Basque region of northern Spain, says Professor Ginder. The Basque were out of favor with the Franco regime following the Spanish Civil War, he notes. Thus, the Basque pooled resources and built their own manufacturing base, initially using area resources. And they built their own cooperative banking system to finance their developments, and their own cooperative colleges and technical schools to keep worker–members skilled with state–of–the–art technologies.

Iowa's resident expert on the Basque model, known as the Mondragon model, is **Gerald Pepper,** the former executive director of the Iowa Institute of Cooperatives. He became a VOCA volunteer (see Chapter 6) after retiring from the Institute in 1987, turning his attention from Iowa co–ops to building local cooperatives in South America. More recently, he's devoted his time and attention to Bulgaria.

Mr. Pepper learned of industrial cooperatives in northern Spain while doing research for his VOCA assignments, he says. "In 25 years, they (the Basque) put together about 100 of these industrial cooperatives. And they've become the leading manufacturer of component parts for stoves, refrigerators and automobiles. They've had enormous success," he says. [9–34] "It occurs to me that there would be a place in rural America where these things would work and thrive through the cooperative system." [9–35]

A form of the Mondragon system is being used by small, rural–based manufacturers in Arkansas, says Christ van Daalen of the Puget Sound Cooperative Federation at Seattle. And his organization is using the concept to promote community development throughout Washington state. Since the Puget Sound office is also one of nine regional cooperative centers being established nationwide, it will also work with neighboring co–op organizations to promote similar developments in Oregon, Idaho, Montana and has plans to reach northwards, into British Columbia.

Forms and variations of Mondragon cooperation have started in central Washington, he says, where local fruit co–ops are pooling resources to share new apple washing equipment. [9–36]. In the

Aberdeen area of southwest Washington, community development groups are working with an agro–forestry co–op to diversify by developing specialty forest product production such as floral greens, medicinal herbs, berries and mushrooms. Another project is establishing a hybrid cottonwood cooperative in which small farmers will raise fast–growing cottonwood trees for use as pulp in nearby paper plants.

And in true Mondragon form, Mr. van Daalen says, Washington cooperative and community development groups are helping form manufacturing networks. One seeks to establish a network of in–home businesses for Hispanic women in the Yakima Valley; two projects, WoodNet and Woodcraft Network, are flexible manufacturing systems to raise value of local forest resources by supplying cabinet makers, local furniture companies and various other woodworking artisans, and a wood co–op is being formed to explore joint venture arrangements with small business, area Native American tribes, non–profit organizations and local governments. "The truth is, we aren't dealing with a spotted owl or salmon problem," he says. "Our whole eco system is in danger. A lot of new direction will come with economic development diversification and value–added strategies. Meanwhile, we have to get our families through the period of transition." [9-37]

ଓଓଓ

Former Agriculture Secretary **Bob Bergland,** president of the National Rural Electric Cooperatives Association form 1981 to 1994, tells of another system of networking involving electric co–ops and embodying the cooperative spirit.

During the farm financial crisis of the mid–1980s, he says, NRECA launched an economic development program whereby the member co–ops and the national organization provided planning and financial support for local manufacturers. "We concluded that we needed to help whoever was keeping our rural communities alive," he says. [9-38] "So we started this program where we'd go into a small business and ask the owner what we could do to help him or her grow and expand.

"Sometimes they would laugh at us. They only had three employees, they would say. And we'd tell them we knew that. But we had come to learn what they needed to grow so they would hire six employees. When they stopped laughing, they started thinking. Some of our best successes didn't make headlines in your big city newspapers, and some didn't make mention in the local, community weeklies. But we have helped create important jobs in a lot of rural towns, and these communities are going to survive!" [9–39]

The NRECA program would help local Chambers of Commerce and city officials do economic assessment studies. There were only three prohibitions built into the system that would make the electric co–ops walk away, Secretary Bergland recalls. One was that the community had to have reasonable access to hospitals and health care facilities. Another was that the community had to have a demonstrated record of supporting their local schools. Public schools and their community education programs are becoming increasingly important for training and retraining workers to use rapidly developing technologies, he says. Finally, rural electrics would walk away from an economic development project if it depended on special tax breaks provided by the community.

"When you start playing games with taxes, your schools are going to pay a price," Secretary Bergland said.. "You probably can't have a good medical care infrastructure, either. You've got to think about what really is community development."

ഇരുന്ന

There is considerable literature in academic circles that explores development issues. Is there really a difference between economic development and community development? Yes, if you approach development using guidelines and benchmarks such as criteria supplied by the Committee for Economic Development. No, if you read the literature of other recognized planning experts who see no difference between the two but argue that business development and community development goals must be managed side by side

I side with the former theorists. Too many communities sell themselves only for jobs.

Lisa Zellmer, a former legislative aide to Senator Kent Conrad of North Dakota, has explored the two concepts in an extraordinary graduate paper while working at the Hubert H. Humphrey Institute's State and Local Policy program. Her employment with the senator occurred during the earlier "second wave" period of development when the purpose behind most activities and assistance was creating jobs. Her research at the Humphrey Institute was during the "third wave" period when leading thinkers advocated "collaboration between government and business, community involvement and capacity and consensus–building." [9–40]

Ms. Zellmer goes beyond the third wave and defines a criteria for community development. She proposes a "reflective leadership" approach for community economic planning that weighs plans against six central ideas. They include (a) quality of life issues, for the entire community; (b) equity issues; (c) institutional or inter–jurisdictional constraints, which require government, business, non–profit institutions and community leaders to work in concert; (d) shared vision, among varied interests groups with the understanding of the community at large; (e) cooperation not competition, in which broader community units work together and not compete for economic development funds or duplicate programs; and (f) fostering innovation, creativity and diversity. [9–41].

Eureka! Ms. Zellmer has found it. Elements for shaping a "fourth wave" development strategy become obvious. On closer look, the difference between community development and economic development isn't a thin line, and, like a U.S. Supreme Court Justice once said about the difference between art and pornography, "I know the difference when I see it." Government and business must be partners in a community development strategy. Where cooperatives now exist, and where they can be formed, they become automatic partners in the community development process.

That message is spreading across Canada these days.

Bill Hatton, president of Development Management Institute at Prince Albert, Saskatchewan, has launched a syndicated newspaper column through the Prince Albert <u>Daily Herald</u> in which he explains concepts of community economic development (CED) that will, in time, be recognized as "fourth wave" development in the United States.

In May, 1994, when he launched his column, Mr. Hatton gave his explanation of CED:

"CED is grass roots economic development. It aims to give back control over their economies to communities and local people. Instead of top–down policies controlled by a bunch of bureaucrats in Ottawa or a provincial capital, CED is a bottoms–up strategy controlled by the folks who have the most to gain from economic development." [9–42]

Indian communities in Canada, and rural areas where a dominant mine industry or fish plant have closed, provide Mr. Hatton with examples of successful CED ventures. In those circumstances, communities have bound together for bootstrap community development if for no other reason than all residents could recognize their common problems. [See Chapter Note B]

But cooperatives in the Prairie Provinces also provide the economic development consultant with successful models. "When people began organizing in the 1920s they were motivated by the high cost of buying the necessities of life or shipping their grain to market," he wrote. "Their prosperity, if not their survival, depended on acting together as a community. As everyone knows, the co–operative movement was very successful." [9–43]

ℰℭℰℭℰℭℰℭ

A new rural America is possible. It can sustain a strong economy and quality of life for its people. In some places, the transformation is starting to take root. It is achievable because there are many models for change to draw from, both at home and from abroad. Judith Sandberg and Stanley Dryer, officials of the Washington, D.C.–based National Cooperative Bank, note the models involve co–ops and related business and service activities found in both rural and urban communities. [9–44]

In a time when much public hand–wringing is spent on American indifference, on our fragmented society, and on our apparently slipping ability to improve the standard of living for a majority of our residents, the National Cooperative Bank reminds us that nearly half of all Americans are involved with cooperative activities every day. Everyday in American, it notes:

- 13.3 million checks with a value of $1.3 billion are cleared by credit unions (which are cooperative institutions).
- 3,000,000 people go home to cooperative housing units.
- 11,000,000 people go to work at 10,000 companies that have employee stock option plans.
- 90,000 people work at jobs that were assisted by cooperative community development corporations.
- 219,000 people have appointments with doctors at cooperative and nonprofit health maintenance organizations.
- $123,000,000 worth of groceries and household products are bought from grocers who are members of wholesale co–ops.
- 60,000 people receive medical care at cooperative community health care centers.
- $1,200,000 worth of goods are bought from natural food cooperatives and buying clubs.
- $1,000,000 worth of hiking, camping and outdoors goods are bought at member–owned Recreational Equipment Incorporated (REI) Stores.
- $1,700,000 in new financing is made available to co–op businesses by the National Cooperative Bank.
- $22,000,000 worth of meals are eaten at restaurants that are supplied by purchasing co–ops.
- $28,500,000 worth of merchandise is purchased at hardware stores supplied by dealer–owned wholesale co-ops.

- 1,800,000 square feet of carpeting is sold by retailers supplied by Carpet One, their wholesale co-op.
- 64,000 people check into hotels that are members of the Best Western International cooperative.
- $219,000,000 in farm products and inputs are sold by agricultural marketing and supply co-ops.
- $148,000,000 in new financing is provided Farm Credit System bank members.
- 50,000,000 people are served by cooperatively owned or affiliated insurance companies.
- 1,200,000 people communicate on telephone lines serviced by telephone co-ops.
- 25,000,000 people receive electricity from consumer–owned electric co-ops.
- 90,000 households receive tv programming beamed to their homes by the National Rural Telecommunications Cooperative.
- $3,200,000 worth of books and merchandise are bought at collegiate bookstore co-ops.
- 7,655 news media outlets share and disseminate worldwide news through the Associated Press cooperative.
- 60,000,000 households access Cspan, a cable tv co-op.
- 50,000 families send children to co-op day–care centers.

And, as there must be an end to everything, and everyday:

- 41 people are buried by member–owned memorial societies. [9–45]

Chapter 9 FOOTNOTES

1. Cantrell, Randy. Associate professor, Hubert H. Humphrey Institute of Public Affairs, University of Minnesota.

2. For the best look at the **Buffalo Commons** arguments put forth by Frank and Deborah Popper, see: Matthews, Anne. Where the Buffalo Roam. Grove Weidenfeld. New York, New York. 1992. Since newspapers and land grant university sociologists and economists had already written widely in the Midwest and Great Plains states of the emergence of ghost towns, the Poppers' major contribution was in attracting greater attention to demographic changes occurring from North Dakota and Montana to west Texas and New Mexico. Unlike similar literature before and since, their argument had a sharp edge: The Poppers said communities should disappear and large areas of land should revert to native grasslands, righting a century of ecological damage caused by agriculture.

3. This is the author's oversimplification of the argument.

4. Vogel, Sarah. Commissioner of Agriculture. State of North Dakota. From interview, August, 1994.

5. Amato, Joseph and Meyer, John W. The Decline of Rural America. Crossings Press. Marshall, Minn. 1993.

6. Norman, Jim. City administrator, Renville, Minnesota. From interview, July, 1994.

7. A shortened version of the following information about Wisconsin, Minnesota and North Dakota new generation cooperatives first appeared in the Business / Twin Cities section of the St. Paul Pioneer Press. Aug. 8, 1994.

8. Olson, Allen I. Head of the Independent Community Bankers of Minnesota and former governor of North Dakota. From interview, July, 1994.

9. DeCramer, Gary. Hubert H. Humphrey Institute of Public Affairs, University of Minnesota. From comments after critiquing Chapter 8, July, 1994.

10. Famalette, Joe. President of American Crystal Sugar. From interview in July, 1994, after his return from visiting with executive of Cebeco, Avebe, Suiker Unie and Rabobank in the Netherlands. [For reference, see Chapter 4.]

11. Cook, Mike. University of Missouri at Columbia. From interviews, and Food and Agricultural Marketing Issues for the 21st Century. The Food and Agricultural Marketing Consortium. 1993.

12. Ibid. Page 159.

13. For perspective on the difficulty co-ops periodically have in adding value to dairy farmers' milk, thus improving farmers' equity through processing and marketing, see: Stern, William M. "Land O' Low Returns." Forbes Magazine. Aug. 15, 1994.

14. New arrangements for partnering with other companies and creating joint ventures or foreign subsidiaries to enter world markets are discussed in Chapters 2 and 5. Dutch cooperatives, as discussed in Chapter 4, are beating American co-ops to the punch in world markets and foreign investments. From visits in the Netherlands, however, the author believes the Dutch head start in global markets may be "defensive," as described by Mike Cook, duplicating the business strategies of their competitors. This suspicion was strengthened in 1994 when Joe Famalette visited Dutch co-ops and said he wasn't convinced the Dutch have determined how their foreign investments will enhance member equity in the co-ops.

15. Greene, Robert. Associated Press. "Small hog producers feel pinch." Farm Forum, a business section supplement of the Aberdeen (S.D.) American-News. July 29, 1994.

16. Ibid.

17. Boehlje, Michael. Purdue University. From interviews and summary published January, 1994, by North Dakota State University Extension Communications.

18. Boehlje, Michael; Schrader, Lee; Foster, Kenneth; and Kadlec, John. "Structure and Coordination." Indiana Agriculture 2000: A Strategic Perspective. Department of Agricultural Economics and Purdue University Cooperative Extension Service, West Lafeyette, Ind. 1992.

19. Cropp, Robert. University of Wisconsin–Madison and Wisconsin Center for Cooperatives. From interviews.

20. Hughes, Harlan. North Dakota State University Extension livestock economist. "The Market Advisor" column distributed Jan., 27, 1994.

21. Hughes, Harlan. "The Market Advisor." Dec. 30, 1993.

22. Fairbairn, Brett. University of Saskatchewan, Saskatoon, Sask. From conversations, July, 1994; and interview, August, 1994.

23. Fairbairn. Letter dated Aug. 26, 1994.

24. Saskatchewan Cooperatives: A Record of Community Development. Centre for the Study of Cooperatives, University of Saskatchewan. Saskatoon, Saskatchewan. 1992.

25. Ibid.

26. Aeilts, David. "Cooperation Par for the Course." AgriVisions Magazine. Harvest States Cooperatives. September/October, 1993.

27. Ibid.

28. Winsor, Susan. "Investing in the Future." Cooperative Profiles. Cenex/Land O'Lakes. Second Quarter, 1993.

29. Ibid.

30. Miller, Patricia. "Carving Out a Niche Market." Cooperative Profiles. Cenex/Land O'Lakes. Second Quarter, 1994.

31. Ginder, Roger. Iowa State University, Ames, Iowa. From interview, August, 1994.

32. West, Garland. "AGRI Industries Members Approve 10 Year Extension of AGM Partnership." Cargill Inc. news release, July 5, 1994.

33. Ibid. Parentheses added by author.

34. Pepper, Gerald. Former executive director, Iowa Institute of Cooperatives, Ames, Iowa. From interview August, 1994.

35. Pepper.

36. Van Daalen, Chris. Puget Sound Cooperative Federation, Seattle, Wash. From interview, August, 1994. For information on Mondragon concepts, he suggests: Morrison, Roy. We Build the Road as We Travel. New Society Publishers. 1991; and Logan, Chris. Mondragon: An Economic Analysis. George Allen and Unwin (London). 1982.

37. Van Daalen. August, 1994.

38. Bergland, Bob. Roseau, Minn. From interview for article: Egerstrom, Lee. "Bergland says farmers must have market before planting." St. Paul Pioneer Press. Aug. 8, 1994.

39. Bergland. From interview.

40. Zellmer, Lisa. "Reflective Leadership in Economic Development." A monograph, unpublished, prepared at the Hubert H. Humphrey Institute of Public Affairs. 1994

41. Ibid.

42. Hatton, Bill. "Communty growth needed to help start up economy." Prince Albert Daily Herald. Prince Albert, Saskatchewan. May 6, 1994.

43. Ibid.

44. Sandberg, Judith; and Dryer, Stanley. From interviews.

45. A Day in the Life of Cooperative America. National Cooperative Bank. Washington, D.C. 1994.

Chapter 9 CHAPTER NOTE 9–A

The following list of 50 "new generation" and community development cooperatives was published along with a locator map of the Upper Midwest states in the St. Paul Pioneer Press on Aug. 8, 1994 ("Where the new co–ops are"). There were at least that many additional ventures being studied, although there was insufficient information on investment or progress to include them in the list.

Sources for the list included Frank Blackburn, Minnesota Association of Cooperatives; Jim Arts, Cooperative Development Service, Madison, Wis.; and Jack Piela, North Dakota Coordinating Council on Cooperatives.

1. **Fargo–Moorhead.** ProGold Co. and Golden Growers Cooperative. Corn processing plant. Site not selected. **$245 million.**

2. **Glenville, Minn.** Forming cooperative to build corn processing plant. **$200 million.**

3. **Marshall, Minn.** Minnesota Corn Processors. Completing expansions and new plant at Columbus, Neb. **$105 million.** Planning beef feedlot in South Dakota to use corn byproducts. **$5.5 million.**

4. **Finley, N.D.** Iso–Straw Cooperative. Forming co–op to supply 180,000 tons of wheat and barley straw to a straw–based particle board factory. **$60 million.**

5. **Jamestown, N.D.** Central Dakota Growers Cooperative forming American Prairie Foods to process potatoes. **$54 million.**

6. **Carrington, N.D.** Dakota Growers Pasta Co. cooperative pasta plant. **$41 million.**

7. **Brookings, S.D.** South Dakota Soybean Processors will build soy processing plant. Site not selected. **$32 million.**

8. **Renville, Minn.** Area farmers are expanding their Co–Op Country grain elevator business, expanding hog production through ValAdCo and Churchill cooperatives, expanding into egg production with Midwest Investors cooperative, and building a joint venture milling business, United Mills. **$31 million.**

9. **Windom, Minn.** Planning an ethanol plant co–op. **$30 million.**

10. **Sleepy Eye, Minn.** Planning an ethanol plant co–op. **$25 million.**

11. **Benson, Minn.** Chippewa Valley Agrifuels Cooperative, building an ethanol plant. **$24 million.**

12. **Little Falls, Minn.** Planning an ethanol plant co–op. **$20 million.**

13. **Washburn, N.D.** McLean County Cattle is forming a cattle slaughter and processing co–op. **$20 million.**

14. **Winthrop, Minn.** Heartland Corn Products Cooperative, building an ethanol plant. **$18.5 million.**

15. **Buffalo Lake, Minn.** Planning an ethanol plant co–op. **$18 million.**

16. **Claremont, Minn.** Planning an ethanol plant co–op. **$18 million.**

17. **Winnebago, Minn.** Corn Plus Cooperative building an ethanol plant. **$17 million.**

18. **Drayton, N.D.** Drayton Grain Processors building a specialty flour mill for wheat–based products. **$17 million.**

19. **Inver Grove Heights, Minn.** Homestead Housing Center has built senior co–op housing projects at Barnesville, Pelican Rapids, Springfield and Wheaton, Minn.; is planning three projects in Iowa and at Rockford, Emily, Grand Marais and Redwood Falls, Minn. **$15 million.**

20. **Granite Falls, Minn.** Forming an alfalfa growers cooperative to supply biomass to planned Northern States Power power plant. **$10–$12 million.**

21. **Edina, Minn.** ReaLife Services. Has eight senior housing co–ops, two more at Rochester, Minn., under development. **$10–$12 million.**

22. **Verndale, Minn.** Verndale Vegetable Growers planning a carrot processing plant. **$10–$12 million.**

23. **Glenwood, Minn.** Planning a vegetable processing cooperative. **$10 million.**

24. **New England, N.D.** Northern Integrated Pork cooperative is building a hog–raising, finishing and processing co–op. **$10 million.**

25. **Antigo, Wis.** Antigo Cheese Co. employees bought Kraft cheese plant by forming a company similar to a workers co–op. Employee investment in plant and business estimated at more than **$4.5 million.**

26. **Clarkfield, Minn.**, and **Engelvale, N.D.** Walton Bean Growers Cooperative is buying two bean processing plants and the name from Walton Bean Co. **$4 million.**

27. **Herman, Minn.** Agassiz Pork Cooperative is building a swine co–op. **$4 million.**

28. **Maddock, N.D.** Maddock Cattle Feeders is building a co–op feedlot. **$3 million.**

29. **LeSueur, Minn.** Cooperative Development Services developed senior co–op housing project. **$2 million.**

30. **Carpio, N.D.** North Central Cattle Feeder Cooperative is building a beef feedlot co–op. **$2 million.**

31. **Walhalla, N.D.** Heartland Feeder Cattle cooperative is building a beef feedlot co–op. **$2.7 million.**

32. **Mayville, N.D.** Heart of the Valley Cooperative is building a dry edible bean processing business. **$2 million.**

33. **New Rockford, N.D.** North American Bison Cooperative is building a bison meat processing and marketing co–op. **$1.6 million.**

34. **Slayton, Minn.** Green Prairie Co–op is building a swine cooperative. **$1.5 million.**

35. **Crosby, N.D.** Quality Pork Cooperative is building a swine co–op. **$1.5 million.**

36. **Hebron, N.D.** Dakota Dairy Specialties co–op is building a specialty cheese factory. **$1.5 million.**

37. **Carrington, N.D.** Northern Plains Organic Co–op is forming an organic milling company. **$1.5 million.**

38. **McClusky, N.D.** Heart of Dakota Pork is building a swine feeding co–op. **$1.5 million.**

39. **Carrington, N.D.** Dakota Aquaculture Cooperative built a fish raising co–op. **$640,000.**

40. **Napolean, N.D.** Napolean Organic Mill cooperative is building a mill for making flours from organically produced grains. **$500,000.**

41. **Hillsboro, N.D.** Planning a carrot raising and processing co–op. Investment costs unknown.

42. **Fergus Falls, Minn.** Western Area Cities Cooperative (WACCO) pools heavy equipment, shares technical employees and buys supplies for 11 western Minnesota cities. No fixed assets.

43. **Sauk Centre, Minn.** An alfalfa co–op is being formed to process and extract juice from alfalfa. Investment costs unknown.

44. **La Farge, Wis.** Coulee Region Organic Produce Producers (CROPP) markets organically grown vegetables and markets organic milk. Investment costs unknown.

45. **Ashland** and **Bayfield, Wis.** Chequamegon Organic Growers forming co–op to market organic fruits, vegetables, meats and processed products

such as salsas, jams, vinegars and salad dressings. Investment costs unknown.

46. **Austin, Minn.** Southern Minnesota Recycling Exchange (SEMREX) formed to coordinate recycling programs for nine area counties. No fixed assets.

47. **St. Paul, Minn.** Minnesota public employees studying a co–op to buy hospital, dental and medical insurance coverage.

48. **Morris, Minn.** Prairieland Products Cooperative forming to market venison products. Investment costs unknown.

49. **Glenburn, N.D.** Glenburn Gardens. Growers are exploring a tomato processing co–op. Investment costs unknown.

50. **Edina, Minn.** United Sugars joint venture formed by transferred assets to market sugars from three Minnesota and North Dakota sugar co–ops, plus future corn sweetener products from ProGold.

Chapter 9 CHAPTER NOTE 9–B

Bill Hatton, president of Development Management Institute at Prince Albert, Saskatchewan, defines the community orientation to economic development in the following column (excerpted) supplied by Dr. Wilbur Maki of the University of Minnesota:

ഇരിഇരിഇരി

Prince Albert Daily Herald   July 21, 1994 (headline:)

"HOME OFFICE HELPS ON LANDING PROJECTS."

"...This column will provide you with the Top 10 basic questions you should ask before you approve or are involved in an economic development project. Just like Dave, we'll start with Number 10.

10. Will the project purchase local goods and services on a preferred basis?

9. Will the project provide extensive training and retraining opportunities for local residents?

8. Will the project provide for quality employment opportunities (like head of household supporting jobs)?

7. Will the project provide for preferred local contracting?

6. Will the project allow for local ownership participation?

5. Will the project address the needs and circumstances of local residents?

4. Will the project require significant re–investment during its lifetime?

3. Will the project be bonded against bankruptcy and failure?

2. Will the project create environmental damage causing long–term problems with nosie, water and air pollution and management of the resulting liabilities?

And now the final Top 10 question from the Home Office in Sioux City, Iowa:

1. Will the project create public infrastructure costs, as well as other public costs, that occur as a result of its development?"

సాసాసా

Mr. Hatton noted that all those questions look at a project from the broader perspective of community, not the single goal of creating jobs. "You must be able to distinguish between a project that brings real, durable benefits to your community and one that 'mines' your community for all it's worth and then leaves for the next community that offers it cut rate taxes or some other inducement," he wrote.

C.T. (Terry) Fredrickson is president and chief executive officer of AgriBank, FCB, and was chairman of the board for the National Council of Farmer Cooperatives in 1994.

AgriBank, based in St. Paul, Minn., was formed by merger of the Farm Credit Banks of St. Paul and St. Louis in 1992. The Farm Credit Bank of Louisville joined in 1994.. It is the largest of the Farm Credit banks, serving an 11 state area of Arkansas, Illinois, Indiana, Kentucky, Michigan, Minnesota, Missouri, North Dakota, Ohio, Tennessee and Wisconsin.

Mr. Fredrickson previously served as chief executive of the Farm Credit Bank of St. Louis from 1986 to 1992 and was the bank's executive vice president and chief operating officer from 1984 to 1986. Before joining the St. Louis bank, he served as Deputy Governor of the Farm Credit Administration in Washington., D.C., held supervisory positions with the Farm Credit Administration, and served as a business analyst with the Wichita Bank for Cooperatives in his native Kansas.

Afterword: C. T. (Terry) Fredrickson

So, where do we go from here? How do you help your community or you local cooperative get started with evaluating its current situation and planning for its future?

There is no one–size–fits–all formula for success in business that will work in rural America, or in large cities, for that matter. Or in New England and the American Southwest. It's doubtful there is a formula that would work in two communities that sit side by side along a highway or rail line. But there is a concept that all communities, and all cooperatives, must share if they are to arrive at a proper course of action. They must become a "strategic organization."

Make No Small Plans offers many examples that support what I'm about to say. I will address my remarks at cooperatives, whether established or new. But they will apply to rural community leaders and planners as well. They will look at the concept of a strategic organization that is willing to confront change, the impediments that cooperatives or communities will encounter trying to create such an organization, and supply critical elements, or "foundation stones" that are necessary for a strategic organization.

THE STRATEGIC ORGANIZATION:

The strategic organization, if it is a business, must be one where consistently superior financial and operating results are driven by its capacity to anticipate changes in its environment. It must anticipate opportunities and threats, and realign itself accordingly. This requires foresight, anticipatory behavior, commitment to superior performance, a strong sense of urgency and superior organizational dexterity. The key element is that the strategic organization is successful because it is strategic.

CONDITIONS UNDER WHICH BEING STRATEGIC IS CRITICAL:

While it may be argued with considerable force that there are few business situations or environments in which being strategic is unimportant, there are some clearly identifiable sets of conditions in which it is critical The five set forth below all seem to rise to that level.

◆Significant or rapid change in the markets in which the organization is engaged. These changes may be manifested in customers, suppliers, competitors or the structure of the markets themselves.

◆New technologies or production methodologies being made available. These are developments which raise the possibility of new ways for the organization to approach its markets.

◆Business environments in which exogenous forces have a very high degree of influence. Examples would include government regulation or involvement, weather, foreign competition or foreign market dependence, and environmental and food safety movements. The more pervasive these factors are, the more important it is for the organization be strategically adept.

◆Highly mature organizations and organizations that operate in a highly mature industry. These organizations have reached a point where they have two options open to them. They may reinvent themselves to extend or renew their life cycle, or they may continue to decline and phase–out their operations.

Since they serve fully or over–served markets, gains may only come from using timing and strategies consistently superior to those of their competition.

◆Businesses which provide homogenous goods and services. Such organizations must rely on being able to formulate and carry out strategies for bringing additional value to the marketplace – or being the low–cost producer. While this is a valid strategy, it is generally regarded as extremely difficult to maintain effectively over time and, therefore, a dangerous approach.

IMPEDIMENTS TO BEING STRATEGIC:

◆Weak and fuzzy definitions.

Most cooperatives have extreme difficulty setting forth a mission statement to serve as the initial element of a strategic organization. In some cases, writing a mission statement to guide the organization becomes a political exercise needing consensus. In other cases, many co–op stockholders and directors are ambivalent about the prospect of financial success.

The result, again, is a mission statement which, when reduced to its essence, says, "We exist to do good things for good people."

◆Misunderstanding of relationships between the cooperative and its customer, patron and owner groups.

Cooperatives must learn to apply proper definitions to three critical terms – patron, owner and customer.

Patrons are a once–a–year phenomenon. They materialize when the cooperative makes the annual distribution of profits, or earnings, as in "patronage distribution."

Owners also make a once–a–year appearance when they choose who will represent their ownership interest as directors. Quite often, these owners immediately become patrons as the checks are passed out after the election is over.

Customers, until recently, has not been a term typically used by cooperatives. The strategically oriented cooperative will be customer driven. Customers shop. The cooperative that premises its existence on member loyalty ignores this reality. It is engaged in the deadly sin of wishful thinking.

The fragmentation of agriculture has caused conflicting customer and owner interests. Traditional agriculture groups tend to see the cooperative as a defensive, cost control mechanism. The commercial group of farmers put less priority on being "protected" from forces in the marketplace.

◆Capital structures are inadequate to support a strategic approach to cooperative direction.

Put in the most elemental terms, the decision to fix something before it is broken is more likely to be viewed as riskier than if the decision is deferred until the nature and magnitude of the precise repair may be determined. The "bet" of the strategic organization is that, over time, enough of its decisions will be right and that the rewards from being right will be more than enough to offset the mistakes.

The higher risks of the strategic organization carry clear implications for capitalization. Its earnings will be more volatile. Its capacity to service debt will be more uncertain. Such an organization must have more capital and, more significantly, a different mix of capital.

The classic cooperative structure of retained earnings was developed in an era when most cooperatives were formed and operated to provide market protection from exploitation. In theory, it offers an effective and equitable method of channeling cooperative earnings to the appropriate patrons. And it provides capital and distributes the burden of that capitalization. But in the real world, it suffers several weaknesses and is quite inadequate to support a strategically oriented organization.

Earnings volatility, including occasional operating losses, is a difficulty for the cooperative that relies on a classic revolving capital plan. It creates equity issues among owners over time. It creates weak to nonexistent means of dealing with operating losses without negative customer–owner reaction.

A base capital plan in which members commit a certain amount of capital for capital improvements and expansions or new ventures addresses some of the deficiencies of the classic revolving plan model. It does not, however, deal with the most vexing aspect of capitalization for the strategic cooperative: From whence does the financial firepower come to take risks if all or most capital is in the form of a future obligation, uncertain only as to timing, payable to the most critical of all the organization's stakeholders?

The inherent limitations imposed on cooperative behavior and performance by the preponderance of cooperative capital structures help explain the infrequency in which cooperatives are market leaders or hold dominate market positions where investment requirements are high, margins and profitability are high over time but highly volatile, and the need to make anticipatory investment is high. Unfortunately, these are the exact circumstances in which cooperatives would have the greatest opportunity to create economic value for their customer–owners.

◆Stockholder value v. Stockholder protection.

The leadership impediment for cooperatives to operate strategically is that far too many boards of directors act as boards of trustees. Their focus is on their fiduciary responsibilities. Preservation of assets, maintenance of proper controls, and assurance of the continued existence of the organization constitute the meets and bounds of their responsibilities. Legal counsel and auditors, internal and external, are their most important resources. The ideal chief executive officer for such a board is an administrative–type officer who will carry out the trusteeship on a day–to–day basis.

It goes without saying that these fiduciary responsibilities are both important and unavoidable, especially in this litigious age. But a strategic business organization will necessarily be directed by a board which recognizes that its most important responsibility is to act as a board of entrepreneurs. This role is to direct or oversee how, where and when corporate resources are to be deployed at risk for the purpose of creating value.

In some respects, this pattern of trustee–type behavior is surprising. Cooperative boards are usually composed of successful members who have demonstrated their own ability to think and respond strategically in their own businesses. They, and those they were elected to represent, would be the obvious beneficiaries of cooperatives focused on creating economic value rather than organizational perpetuation.

On the other hand, the board of trustees approach is consistent with the more traditional, reactionary, defensive concept of the cooperative. If one views the cooperative exclusively as a means of protecting against market imperfections – "keeping the big boys honest" – risk avoidance, asset preservation and organizational survival are the key priorities. This is based on the historic view of the cooperative as a cost center, representing the cost of protection from imperfections in the market.

Foundation Stones of the Strategic Cooperative:

Leaders who would make their cooperatives a strategic organization will be governed by their own history and circumstances. But there are a few elements, or foundation stones, that must be taken into account.

◆First, all confusion regarding the mission or purpose of the cooperative must be eliminated. The mission statement of the strategic cooperative must establish that it is a business existing and operating to increase the profitability of its owners through creating economic value. Nothing more and nothing less.

◆Second, the organization must answer the question: How will we measure our success? This need must be handled through a demanding, objectively measurable, business performance–based definition of success with a scoring system based directly on it.

How this is structured will depend on the nature of the business involved. But it should define success by setting measures for financial and operating performance, efficiency, market penetration and customer satisfaction. And individual measures must be at or above general industry levels of performance if the cooperative's objective is to generate economic value.

◆Third, the organization must make an absolute commitment to be market and customer driven. The capacity to understand, evaluate, quantify and respond to markets better and faster than competitors must be the ultimate objective.

◆Fourth, success as a strategic organization must proceed from a position of strength and competence in the company's core business or businesses. The organization which lacks the capacity to bring superior value to customers in its areas of strength and specialty cannot be strategically successful.

♦Fifth, the organization must create an internal culture which fosters risk assessment, assumption and management rather than risk avoidance, change rather than stability, intensity and improvement rather than complacency, incentive rather than traditional compensation.

♦Finally, the organization must adopt and implement a capitalization program which recognizes that its capacity to assume risk and create value is directly related to the amount and character of its capital resources

ACKNOWLEDGMENTS

It is fitting that this book about reviving our agrarian economy – our farms and rural communities – owes much to good lunches and dinners. If it has a starting point, it would be a series of lunches at the Cedar Street Cafe in Saint Paul, Minnesota, with Gary DeCramer, a fellow of the Hubert H. Humphrey Institute of Public Affairs at the University of Minnesota, and Tom Sand, a research assistant for the Minnesota State Senate.

Mr. DeCramer and I had both participated in sabbatical programs provided by the Atlantic–Pacific Exchange Program that is based in Rotterdam, the Netherlands. We had seen Dutch and other European farm groups and their cooperatives making adjustments to changing world markets, and we wished American farm leaders were aware of their European cousins' creativity. Mr. Sand, meanwhile, served as a special assistant to Agriculture Secretary Bob Bergland and had visited farms and agribusinesses on all of the world's populated continents. He, too, knew that food production and marketing knowledge aren't exclusive properties of American agriculture.

The three of us began encouraging Midwest cooperative leaders to visit Europe and study what American producer–owned businesses might do to expand their operations and bring more value–added processing to their communities. This led to more lunches and occasional dinners with an expanding circle of friends who are all searching for ways to position their communities and businesses for the Twenty–First Century.

Allen Gerber, executive director of the Minnesota Association of Cooperatives, and William J. Nelson, president of The Cooperative Foundation and the cooperative education specialist for Cenex, were most supportive and encouraging as this book grew in both size and objective. It was first envisioned as a manual for cooperative action. Lunches, conferences, dinners and news events prompted people to offer ideas and suggestions for the book. The book grew.

So it is, as with most nonfiction books, there are people who must be thanked because without them this book would not exist. They include Richard Magnuson, a cooperative law expert at Doherty Rumble and Butler; Allen I. Olson, the former governor of North Dakota who heads the Independent Community Bankers of Minnesota; John O'Day, vice president of AgriBank; Lani Jordan and her colleagues at Cenex, and its joint venture, Cenex/Land O'Lakes Ag Services; Jim Erickson at Harvest States Cooperatives and Terry Nagle at Land O'Lakes, two outstanding sources of information as well as gifted writers whose suggestions I tried to heed; Myron Just, a former North Dakota Commissioner of Agriculture and board member of Harvest States; Nellie Mae Crank and John Hendel, dedicated former employees at Farmland Industries; Dana Persson, president and manager of Co–Op Country Elevators at Renville, Minnesota; Robert Williams, Lia Rosenbrand, and Dolly Lim, in Rotterdam, and Joan Canty in Seattle, executives of the Atlantic–Pacific Exchange Program;

T.J. Gilles, congressional aide and former journalist at Great Falls, Montana; Steve and Diana Carney, farmers and community leaders at Peerless, Montana, from where Steve is a Harvest States board member; Frank Blackburn, education director at the Minnesota Association of Cooperatives; Ed Slettom, past executive of the Minnesota Association of Cooperatives; David Birkholz, a brilliant educator and friend who keeps me thinking about global events; Ella Krucoff at the European Union office in Washington and Harald von Witzke, now at Humboldt University in Berlin, who keep me aware of changes in Europe; and James Abourezk, a South Dakota attorney and former U.S. senator who has opened doors for me on four continents.

Also, Donal Heffernan, director of international contract law at Cray Research; Mark Hanson, a cooperative and environmental attorney at Doherty Rumble and Butler; food industry securities analysts George Dahlman at Piper Jaffray, Bonnie Wittenburg at Dain Bosworth and Craig Carver at John G. Kinnard; Garland West at Cargill; Craig Shulstad at General Mills; Terry Thompson and Nancy Perron at Pillsbury; Mark Witmer at Michael Foods; Joe Famellete at American Crystal and Mark Dillon at ProGold and Golden Growers; Alan Drattell, The World Bank; John and Roxanne Aschittino, St. Paul librarians whose friendship was tapped for information and encouragement; Tom Cochrane, executive director of the Minnesota Agri–Growth Council; Stuart S. Peterson, senior vice president and chief credit officer, St. Paul Bank (for Cooperatives); Judith Sandberg, vice president, corporate banking, National Cooperative Bank; Al Christopherson, president of the Minnesota Farm Bureau Federation; and Dave Frederickson, president of the Minnesota Farmers Union.

There are also academics – economists, political scientists, social scientists, plant scientists and professors of literature, English and journalism – who also deserve my thanks. I've cribbed their knowledge, ideas and research in my work as a journalist for the past three decades, and they have shaped my current thoughts.

They include Robert Becker and James Davis (formerly) of St. Cloud State University; Joe Amato, Dave Pichaske, Bill Holm and Leo

Dangel, Southwest State University; Robert Cropp and Ed Jesse, University of Wisconsin at Madison; Dean Gary Rohde, University of Wisconsin at River Falls; Duane Dailey, Mike Cook and Harold Breimyer, University of Missouri at Columbia; President Jon Wefald, Kansas State University; Mike Boehlje, Purdue University; Kathleen Davis, Texas A & M University; Neil Harl, Roger Ginder and Clifford LaMotte, Iowa State University; and the resource pool of gifted and helpful thinkers at the nearby University of Minnesota: Dean G. Edward Schuh, C. Ford Runge, Vernon Ruttan, Steve Taff, Willis Peterson, Willard Cochrane, Jim Houck, Randy Cantrell, Bert Sundquist, Earl Fuller, Richard Levins and Dani O'Reilly, to name only a few.

Economist Wilbur Maki at the University of Minnesota and historian Brett Fairbairn at the University of Saskatchewan deserve special thanks for critiquing key chapters and directing me to available research and development literature. I have long admired Professor Maki and his work on U.S. and regional development issues; I have only recently had the privilege of knowing Professor Fairbairn although I have been familiar with the work of his brother, Garry Fairbairn, who is a leading agricultural journalist in Canada.

There are also current and former journalists, including editors and publishers, who have contributed to this book. Some of them gave assignments that steered me into food, agriculture, foreign trade and rural affairs writing; others inspired me through their work; and I am indebted to others because they had the patience to read and edit the manuscript or encouraged me to finish this book.

It is a disservice to them, but I will simply list them and their current or germane past affiliations: Ray Howe, publisher, Lone Oak Press; William Broom, Al Eisele, Ed Zuckerman, Gil Bailey and the late Walter T. Ridder, Ridder Publications and Knight–Ridder Newspapers; Juan Miquel Pedraza, AgWeek Magazine; Audrey Mackiewicz, National Association of Agricultural Journalists, Sandusky, Ohio; Don Muhm, Dirck Steimel, George Anthan and Jerry Perkins, Des Moines Register; Dan Looker, Successful Farming; Harold Higgins, Boulder Daily Camera; Richard Orr, Chicago Tribune; Don Kendall and Gene Lahammer, Associated Press; Wayne Falda, South Bend Tribune; Bernie Brenner and Drew von Bergen, United Press International; Sharon Schmickle and Frank Wright, Minneapolis Star Tribune; Claudia Waterloo, numerous publications; Michael Flaherty, Wisconsin State Journal; Paul Adams and Gary Gunderson, Agri News; Linda Vance, Linda Kendall and Bob Denman, Commodity News Service (now Knight–Ridder Financial);

Sonja Hillgren, Marcia Taylor, Karen Freiberg, Mary Thompson and Patricia Klintberg, Farm Journal; Bruce Ingersoll and Scott Kilman, Wall Street Journal; Kay Ledbetter, Amarillo Globe News; Steven Lee, Dallas Morning News; Ron DeChristopher and Myron Williams, Iowa Farmer Today; Dan Miller, Progressive Farmer; June Sekoll, Culpeper Farm Chronicle; John Peterson, Detroit News; Mikkel Pates, Fargo Forum; Ann Toner–Gottwald, Omaha World Herald; Steve Painter, Wichita Eagle; Roger Runningen, Nichols and Dezehall Communications; and Jim Webster, Webster Communications;

John and Jan Almen, Kerkhoven Banner; Del Griffin, Aberdeen American–News; Jack Haggerty, Mike Maidenberg and Jim Durkin, Grand Forks Herald; Gene Myers, Kansas City Star; Tom Daly, Duluth News–Tribune; Larry LaMotte and Greg LaMotte, both with CNN; Axel Krause, the International Herald Tribune writer in Paris who inspired me to write this book by telling me his plans to write Inside the New Europe; Ernie Wilkinson, Indianapolis Star; Carl West, Frankfurt State Journal; Carol James at Carol James Communications, Washington, D.C.; former editors John Finnegan, Don O'Grady, Bill Cento, Harry Burnham, Fred Heaberlin and Tom Matthews, and a few hundred current colleagues at the St. Paul Pioneer Press.

Among the latter are Clark Morphew, the religion writer who read the manuscript as an "outsider" unfamiliar with the jargon of agriculture, farm policies and development economics; and Judith Willis, the deputy business editor who allowed me to combine research for the final chapter with features for her pages.

Most importantly, I must thank the four women in my life, Lalinda LaMotte Egerstrom and our daughters Marisa, Kirsten and Kari. We now appreciate why authors include family members in their acknowledgments.

Lee Egerstrom   Maplewood, Minnesota, November, 1994

Bibliography

For Further Reading:

•Abourezk, James G. Advise & Dissent: Memoirs of South Dakota and the U.S. Senate. Lawrence Hill Books. Chicago, Ill. 1989.

•Agricultural Biotechnology: The Next "Green Revolution"? Agriculture & Rural Development Department, The World Bank / Australian Centre for International Africultural Research / Australian International Development Assistance Bureau / International Service for National Agricultural Research. Technical Paper No. 133, 1991. Stock No. 11741, ISBN 0–8213–1741–5.

•Agricultural Extension: The Training & Visit System. Daniel Benor, James Q. Harrison and Michael Baxter. 1984, The World Bank. 95 pgs. Stock No. 10140 (English) 10902 (French)

•Agricultural Extension for Women Farmers, Katherine A. Saito and C. Jean Weidemann. Discussion Paper No. 103, 1990, The World Bank, 74 pgs. Stock No. 11657, ISBN 0–8213–1657–5

•Agricultural Marketing: The World Bank's Experience. A World Bank Operations Evaluation Study, 1990 102 pgs. Stock No. 11535

•Agricultural Mechanization: Issues & Options. A World Bank Policy Study 1987, 96 pgs. Stock No. 10903

•Agricultural Policies, Markets and Trade. Organization for Economic Cooperation and Development. Paris, France. 1989.

•The Agricultural Situation in the Community. Commission of the European Communities. Office of Official Publications. Luxembourg. 1984 Report. 1990 Report.

•Agricultural Technologies for Market–Led Development Opportunities in the 1990s. Ed. Shawki Barghouti, E. Cromwell, A.J. Pritchard. Technical Paper No. 204, 1993 180 pgs Stock No. 12462 ISBN 0–8213–2462–4

•Agricultural Trade Liberalization: Implications for Developing Countries. Ed. Ian Goldin, Odin Knudsen. Copublished by The World Bank & The Organsiation for Economic Co–operation & Development. 1990. 388 pgs. Stock Nos. 11527(English), 11528(French)

•Agricultural Transition in Central & Eastern Europe & The Former U.S.S.R. Ed. Avishay Braverman, Karen M. Brooks & Casaba Csaki. A World Bank Symposium (Proceedings of a conference convened to examine how the region is making the transition to a new agrarian system based on private ownership, true cooperatives, and a market economy. 1992 336 pgs. Stock No. 12322 ISBN 0–8213–2322–9

•<u>Agriculture and Environmental Challenges</u>: Proceedings of the 13th Agricultural Sector Symposium. Ed. J.P. Srivastava, H. Alderman, 1993, 302 pgs. The World Bank, ISBN 0–8213–2585–X

•<u>Agroindustrial Project Analysis: Critical Design Factors</u>. James E. Austin, Johns Hopkins University Press, 2nd Edition 1992. 272 pgs. ISBN 0–8018–4530–0. World Bank Stock No. 44530

• Amato, Joseph A. <u>Servants of the Land</u>. ("The Trinity of Belgian economic folkways in southwestern Minnesota.") Crossings Press. 1990.

•Amato, Joseph and Meyer, John W. <u>The Decline of Rural Minnesota</u>. Crossings Press. Marshall, Minn. 1993.

•<u>American Cooperation 1994</u>. National Council of Farmer Cooperatives. Washington, D.C. 1994.

•Berry, Wendell. <u>Home Economics</u>. North Point Press. Berkeley, Calif. 1987.

•Blei, Norbert. <u>Meditations on a Small Lake</u>. Ellis Press. Peoria, Ill. (Now at Granite Falls, Minn.) 1987.

•Bond, Alec. <u>Phebus Lane</u>. Spoon River Poetry Press. Peoria, Ill., and Granite Falls, Minn. 1987.

•Broehl, Wayne G. Jr. <u>Cargill: Trading the World's Grain</u>. University Press of New England. Hanover, N.H. 1992.

•Brown, Lester R. <u>World Without Borders</u>. Random House. New York, N.Y. 1972.

•Childs, Marquis. <u>Sweden: The Middle Way</u>. Out of print. Several editions were published in the 1950s and 1960s; copies may be found in major research libraries.

•Childs. <u>Sweden: The Middle Way On Trial</u>. Yale University Press. New Haven, Conn. 1980.

•Clark, Thomas D. <u>Frontier America: The Story of the Westward Movement</u>. Charles Scribner's Sons. New York. N.Y. 1959.

•Cochrane, Willard. <u>Anatomy of an Agricultural Crisis</u>. Rowman & Littlefield Publishers. Lantham, Md. 1992.

•Colchester, Nicholas; and Buchan,David. <u>Europower</u>. The Economist Books Ltd. London, Great Britain and Random House, New York, N.Y. 1990.

•<u>A Common Agricultural Policy for the 1990s</u>. Office of Official Publications of the European Communities. Luxembourg. 1991.

•Crossley–Holland, Kevin. <u>The Norse Myths</u>. Pantheon Books. New York, N.Y. 1980.

•Curtis, Natalie. <u>The Indians' Book</u>. Bonanza Books. New York, N.Y. 1987.

•Doyle, Jack. <u>Altered Harvest</u>. Viking. New York, N.Y. 1985.

•Drache, Hiram M. Koochiching. Interstate Printers & Publishers. Danville, Ill. 1983.

•Dracke, Hiram M. Plowshares to Printouts. Interstate Printers & Publishers. Danville, Ill. 1985.

•Eisele, Albert. Almost to the Presidency: A Biography of Two American Politicians. The Piper Co. Blue Earth, Minn., and San Francisco, Calif. 1972.

•Erb, Guy F. and Kallab, Valeriana. (Editors) Beyond Dependency. Overseas Development Council. Washington, D.C. 1975.

•Etter, Dave. Alliance, Illinois. Spoon River Poetry Press. Granite Falls, Minn. 1983.

•Etter. Selected Poems. Spoon River Poetry Press. 1987.

•Etter. Sunflower County. Spoon River Poetry Press. 1994.

•Facts and Figures 1992. Ministry of Agriculture, Nature Management and Fisheries. The Hague, the Netherlands. 1992.

•Fairbairn, Brett; Bold, June; Fulton, Murray; Ketilson, Lou Hammond; and Ish, Daniel. Cooperatives & Community Development. Centre for the Study of Cooperatives. University of Saskatchewan, Saskatoon, Sash. 1991

•Fite, Gilbert C. Beyond the Fence Rows: A History of Farmland Industries. University of Missouri Press. Columbia, Mo. 1978.

•Freivalds, John. The Famine Plot. Stein and Day. New York, N.Y. 1978.

•Freivalds, John. Grain Trade: The Key to World Power and Human Survival. Stein and Day. New York. N.Y. 1976.

•Galbraith, John Kenneth. Economics & the Public Purpose. Houghton Mifflin Co. Boston, Mass. 1973.

•Gartrell, David, and Henderson, Byron. (Editors) Globalization and Relevance of Cooperatives. Centre for the Studies of Cooperatives, University of Saskatchewan. Saskatoon, Sask. 1991.

•Gilmore, Richard.. A Poor Harvest The Clash of Policies and Interests in the Grain Trade. Longman Inc. New York, N.Y. 1982

•Hartling, Poul. The Danish Church. Det Danske Selskab. Copenhagen, Denmark. 1964.

•Hasselstrom, Linda. Windbreak: A Woman Rancher on the Northern Plains. Barn Owl Books. Berkeley, Calif. 1987.

•Hassler, Jon. Grand Opening. Ballantine Books. New York, N.Y. 1987.

•Hassler, Jon. Staggerford. Ballantine Books. New York, N.Y. 1974.

•Heffernan, Donal. Hillsides. Lone Oak Press. Rochester, Minn. 1990.

•Heffernan. Orion. Lone Oak Press. Rochester, Minn. 1993.

•Herbers, John. The New Heartland. Times Books. New York, N.Y. 1978.

•Hill, Samuel S. Jr. Religion and the Solid South. Abington Press. Nashville, Tenn. 1972.

•Hofstadter, Richard. The Age of Reform. (Paperback) Vintage Books. New York, N.Y. 1955.

•Hollerman, Leon. Japan, Disincorporated: The Economic Liberalization Process. Hoover Institution, Stanford University. Stanford, Calif. 1988.

•Holm, Bill. Coming Home Crazy. Milkweed Editions. St. Paul, Minn. 1991.

•Holm. The Dead Get By With Everything. Milkweed Editions. St. Paul.. 1993.

•Howard, Robert West. The Vanishing Land. Ballantine Books. New York. N.Y. 1985.

•Johnson, D. Gale and Schuh, G. Edward. The Role of Markets in the World Food Economy. Westview Press. Boulder, Colo. 1983.

•Kirmmse, Bruce H. Kierkegaard in Golden Age Denmark. Indiana University Press. Bloomington, Ind. 1990

•Kohn, Howard. The Last Farmer: An American Memoir. Harper & Row. New York, N.Y. 1988.

•Kopytoff, Igor. (Editor) The African Frontier: The Reproduction of Traditional African Societies. Indiana University Press. Bloomington, Ind. 1989.

•Korsching, Peter; Borich, Timothy; and Steward, Julie. (Editors.) Multicommunity Collaboration: An Evolving Rural Revitalization Strategy. North Central Regional Center for Rural Development. Iowa State University. Ames, Iowa. 1992.

•Krueger, Anne O.; Michalopoulos, Constantine; and Ruttan, Vernon W. Aid and Development. The Johns Hopkins University Press. Baltimore, Md. 1989.

•Legrand, Jacques. Cut in Antwerp. HRD. Antwerp, Belgium. 1982.

•Lewis, Finlay. Mondale. Harper & Row. New York, N.Y. 1980.

•Lewis, John P. and Kallab, Valeriana. (Editors) Development Strategies Reconsidered. Overseas Development Council. Washington, D.C. 1986.

•Liberman, E.G. Economic Methods and the Effectiveness of Production: A Study of Soviet Economic Reforms. Ekonomika Publishing House. Moscow, Russia. 1970; and Anchor Books. New York, N.Y. 1973.

•Logan, Chris. Mondragon: An Economic Analysis. George Allen and Unwin. London, U.K. 1982.

•MacAvoy, Paul W. (Editor) Federal Milk Marketing Orders and Price Supports. American Enterprise Institute. Washington, D.C. 1977.

•Making Development Sustainable. The World Bank. Washington, D.C. 1994.

•Manfred, Frederick. Duke's Mixture. The Center for Western Studies. Augustana College, Sioux Falls, S.D. 1994.

•Manfred.. No Fun on Sunday University of Oklahoma Press. Norman, Okla. 1990.

•Martin, Albro. James J. Hill & the Opening of the Northwest. Oxford University Press. New York, N.Y. 1976.

•Matthews, Anne. Where the Buffalo Roam. Grove Weidenfeld. New York, N.Y. 1992.

•McCarthy, Eugene J. And Time Began. Lone Oak Press. Rochester, Minn. 1993.

•McCarthy. The Limits of Power. Holt, Rinehart and Winston. New York, N.Y. 1967.

•McGovern, George. Grassroots. Random House. New York, N.Y. 1977.

•Meissner, Frank. Seeds of Change: Stories of IDB Innovation in Latin America Dr. The Johns Hopkins University Press. Washington, D.C. 1991.

•Moberg, Vilhelm. The Immigrants. (Paperback) Popular Library and Simon & Schuster. New York, N.Y. 1951.

•Moe, Richard. The Last Full Measure. Henry Holt and Co. New York. N.Y. 1993.

•Morrison, Roy. We Build the Road as We Travel. New Society Publishers. 1991.

•Nass, David. Holiday: Minnesotans Remember the Farmers' Holiday Association. Plains Press. Southwest State University, Marshall, Minn. 1984.

•Norris, Kathleen. Dakota: A Spiritual Geography. Tichnor & Fields. New York, N.Y. 1993.

•Osborne, David. Economic Competitiveness: The States Take the Lead. Economic Policy Institute. Washington, D.C. 1987.

•Owensby, Walter L. Economics for Prophets. Wm. B. Eerdmans Publishing Co. Grand Rapids, Mich. 1988.

•Paarlberg, Robert L. Fixing Farm Trade: Policy Options for the United States. Ballinger Publishing Co. Cambridge, Mass. 1988.

•Paddock, Joe; Paddock, Nancy; and Bly, Carol. Soil and Survival. Sierra Club Books. San Francisco, Calif. 1986.

•Peoples, Kenneth L.; Freshwater, David; et. al. Anatomy of an American Agricultural Credit Crisis: Farm Debt in the 1980s. Rowman & Littlefield Publishers. Lanham, Md. 1992.

•Pichaske, David. <u>Poland in Transition</u>. 1989 – 1991. Ellis Press. Granite Falls, Minn. 1994.

•Powers, J.F. <u>Wheat That Springeth Green</u>. Alfred A. Knopf. New York, N.Y. 1988.

•Prestwich, Roger and Taylor, Peter. <u>Introduction to Regional &</u> <u>Urban Planning in the United Kingdom</u>, Longman Harlow, United Kingdom, 1990

•Radzilowski, John. <u>Out on the Wind: Poles and Danes in Lincoln County, Minnesota, 1880–1905</u>. Crossings Press. Marshall, Minn. 1992.

•Riegel, Robert E. and Athearn, Robert G. <u>America Moves West</u>. (Fourth Edition.) Holt, Rinehart and Winston. New York, N.Y. 1966.

•Rolvaag, O.E. <u>Giants in the Earth</u>. (Paperback) Harper & Row. New York, N.Y. Perennial Classics edition of the 1927 novel.

•Runge, C. Ford. (Editor) <u>The Future of the North American Granary: Politics, Economics, and Resource Constraints in North American Agriculture</u>. Iowa State University Press. Ames, Iowa. 1986.

•Ruttan, Vernon. <u>Agriculture Research Policy</u>. University of Minnesota Press. Minneapolis, Minn. 1982.

•Ruttan, Vernon. <u>Why Food Aid?</u> Johns Hopkins University Press. Baltimore, Md. 1992.

•Schertz, Lyle P. and Daft, Lynn M. (Editors) <u>Food and Agricultural Markets: The Quiet Revolution</u>. National Planning Association. 1994.

•Schwab, Jim. <u>Raising Less Corn and More Hell</u>. The University of Illinois Press. Champaign, Ill. 1988.

•Senauer, Ben; Asp, Elaine; and Kinsey, Jean. <u>Food Trends and the Changing Consumer</u>. Eagan Press. St. Paul, Minn. 1991.

•Severson, Harlan M. <u>Stepping Forward, Boldly: The Story of East River Electric Power Cooperative</u>. Hunter Publishing. Madison, S.D. 1975.

•Simon, Julian L. <u>The Economic Consequences of Immigration.</u> Basil Blackwell Inc. Cambridge, Mass. 1989.

•Smiley, Jane. <u>A Thousand Acres</u>. Fawcett Columbine. New York, N.Y. 1991.

•Stavenhagen, Rodolfo. <u>Social Classes in Agrarian Societies</u>. Anchor Press/Doubleday. Garden City, N.Y. 1975.

•Stuhler, Barbara. <u>Ten Men of Minnesota and American Foreign Policy</u>. Minnesota Historical Society. St. Paul, Minn. 1973.

•Sunbury, Ben. <u>The Fall of the Farm Credit Empire</u>. Iowa State University Press. Ames, Iowa. 1988.

•Vinz, Mark and Tammaro, Thom. (Editors) Common Ground: A Gathering of Poets from the 1986 Marshall Festival. Dacotah Territory Press. Moorhead, Minn. 1988.

•Wallace, Henry A.; and Brown, William L. Corn and its Early Fathers. Iowa State University Press. Ames, Iowa. Revised Edition. 1988.

•Weaver, Will. A Gravestone Made of Wheat. Simon & Schuster. New York, N.Y. 1989.

•Weaver. Red Earth, White Earth. Simon & Schuster. New York, N.Y. 1986.

•Wilkins, Robert P. and Wynona H. North Dakota. A Bicentennial History. W.W. Norton & Co. New York, N.Y. 1977.

•Willard, Helen. Pow–Wow and Other Yakima Indian Traditions. Roza Run Publishing. Prosser, Wash. 1990.

•Development and the Environment., World Development Report 1992, The World Bank, Washington, D.C. 1992.

•Yenne, Bill. The Encyclopedia of North America Indian Tribes. Arch Cape Press. Greenwich, Conn. 1986.

•Zuckerman, Edward. Almanac of Federal PACs. 1986. Amward Publications. Washington, D.C. 1986.

Magazines, Monographs, Newspaper articles

•"ACDI Signs First Protocol with Mexico." Cooperative News International. Agricultural Cooperative Development International. Washington, D.C. Volume 6, No. 4. 1993.

•Adams, Paul. "Bill would help deal with feedlot waste." Agri News. (Weekly supplement of the Rochester Post–Bulletin. Rochester, Minn. March 24, 1994.

•Adams. "Group alleges 66 new feelot rules violations." Agri News. March 24, 1994.

•Adams. "MPCA backs off on feedlot rule change." Agri News. Oct. 7, 1993.

•Adams. "River protection focuses on feedlots." Agri News. July 1, 1993.

•Adams. "State moves to reclaim namesake river." Agri News. July 1, 1993.

•Aeilts, David. "Cooperation Par for the Course." AgriVisions Magazine. Harvest States Cooperatives. St. Paul, Minn. September/October, 1993.

•"Agricultural Policy Council Meeting Discusses Future of Japanese Agriculture." And, "ZEN–CHU (Central Union of Agricultural Cooperatives) Files Demand with Ministry of Agriculture,

Forestry and Fisheries Minister to Amend Food Control System." Japan Agrinfo Newsletter. August, 1993.

•"Agro–Industry." Denmark Presents. Export Promotion Danmark. Copenhagen, Denmark. June, 1993.

•Barr, Terry N. "Cooperatives in the 21st Century and Implications for the Farm Credit System." National Council of Farmer Cooperatives. Washington, D.C. 1993.

•Beaver, Stephen. "Dairy 2020 Area Councils, Working Group Explore Farm Management Tools." Dairy 2020 Update newsletter. University of Wisconsin–Madison and Wisconsin Department of Agriculture Trade and Consumer Protection. Madison, Wis. Aug. 17, 1993.

•Beaver, Stephen. "Dairy 2020 Looks Into Entry/Exit/Transfer of Dairy Farms." Dairy 2020 Update. Aug. 17, 1993.

•Beal, Dave. "Three states vying for sweet business." St. Paul Pioneer Press. St. Paul, Minn. June 24, 1994.

•Bourne, Jana. "Pumping New Life into Rural Communities." Cooperative Partners. Cenex/Land O'Lakes Ag Services. Inver Grove Heights, Minn. March–April, 1994.

•Campbell, Will D. "The Federation of Southern Cooperatives: Staying the Course." World Magazine. The Journal of the Unitarian Universalist Association. September/October 1994.

•Cassel, Andrew. "Buffalo Commons Thesis Angers N. Dakotans." Knight–Ridder News Service. (Datelined: Bismarck, N.D.) Nov. 15, 1989.

•Chapman, Dan. "Behind the bid for Mercedes." Charlotte Observer. Charlotte, N.C. Oct. 11, 1993.

•Cobb, Kathy. "Sioux Falls can bank on (John) Morrell – for at least 10 years." fedgazette. Federal Reserve Bank of Minneapolis. Minneapolis, Minn. July, 1994.

•Cook, Michael. Research paper printed in: Food and Agricultural Marketing Issues for the 21st Century. Compiled by the Food and Agricultural Marketing Consortium, University of Missouri. Columbia, Mo. 1993.

•Cooperative Profiles. (Quarterly for directors of Cenex / Land O'Lakes Cooperatives. First, Second, Third and Fourth Quarters, 1993 and 1994. Cenex / Land O'Lakes. St. Paul, Minn.

•A Day in the Life of Cooperative America. (Promotional brochure.) National Cooperative Bank. Washington, D.C. 1994.

•Dawson, Jim. "Russian farm project to have local support." Star Tribune. Minneapolis, Minn. Jan. 15, 1993.

•DeCramer, Gary. (Untitled sabbatical report.) Atlantic–Pacific Exchange Program. Rotterdam, the Netherlands; and Hubert H.

Humphrey Institute of Public Affairs, University of Minnesota. Minneapolis, Minn. 1989.

•"Drayton Grain Processors exceed requirements." Fargo Forum. Fargo, N.D. June 21, 1994.

•Egerstrom, Lee. "Agriculture's Flight of Capital." (Series) St. Paul Pioneer Press. St. Paul, Minn. Beginning April 7, 1986.

•Egerstrom. "Keeping Things Sweet at Crystal Sugar." St. Paul Pioneer Press. April 11, 1994.

•Egerstrom. "Minnesotans lose naivete by setting limits for themselves." St. Paul Pioneer Press. Dec. 21, 1987.

•Egerstrom. "A New Lease on Life." St. Paul Pioneer Press. Jan. 9, 1994.

•Egerstrom. Newspaper Farm Editors in a Changing World. (Monograph) Newspaper Farm Editors of America. Sandusky, Ohio. April, 1987.

•Egerstrom. "Prince Ties Oil Tab, U.S. Profits." Pasadena Star–News. Pasadena, Calif. Jan. 13, 1974.

•Egerstrom. Rediscovering Cooperation. (Monograph) Regional Director Workshop. Minnesota Association of Cooperatives. St. Paul, Minn. January, 1993.

•Egerstrom. "The Search for Greener Pastures." St. Paul Pioneer Press. July 25, 1993.

•Egerstrom. "Troubled Farms." (Series) St. Paul Pioneer Press. Beginning, Jan. 21, 1985.

•Fairbairn, Brett. "Cooperatives and Globalization: Market–driven Change and the Origins of Co–operatives in the Nineteenth and Twentieth Centuries." An essay paper presented at the 1991 Meetings, Canadian Association for Studies in Co–operation. Queen's University; and published in Globalization and Relevance of Co–operatives. Centre for the Study of Co–operatives, University of Saskatchewan. Saskatoon, Sask. 1991.

•"Fine Foods, Functional Foods, and the Research that has made Denmark an International Food Force." (A package of articles.) Denmark Review. Royal Danish Ministry of Foreign Affairs, Secretariat for Foreign Trade. Copenhagen, Denmark. February, 1994.

•"Food and Agricultural Policy Reforms in the Former USSR: An Agenda for the Transition." Studies of Economies in Transformation. The World Bank. Washington, D.C. September, 1992.

•Foreign Agricultural Trade of the United States (FATUS). Economic Research Service, U.S. Department of Agriculture. Washington, D.C. Annual reports, 1991, 1992, 1993.

•Former USSR. International Agriculture and Trade Reports. U.S. Department of Agriculture, Economic Research Service. Washington, D.C. May, 1993.

•Glynn, Priscilla. "Job Training Program Performs Better in Rural Areas." Farmline. U.S. Department of Agriculture, Economic Research Service. Washington, D.C. February, 1993.

•Perspectives on U.S. and Canadian Agriculture Government Policies and Great Lakes Shipping and Maritime Policies. University of Minnesota – Duluth. Duluth, Minn. April, 1988.

•Greene, Robert. "Small hog producers feel pinch." Associated Press dispatch, Farm Forum. Business supplement published by Aberdeen American–News, Aberdeen, S.D. July 29, 1994.

•"Growing by Joint Venture." St. Paul Pioneer Press. St. Paul, Minn. Oct. 18, 1993.

•Hansen, Otto Ditlev and Jorgensen, Erik Juul. Food Technology in a Modern Food System – The Case of Denmark. Institute for Food Studies & Agroindustrial Development. Copenhagen, Denmark. 1993.

•Hatton, Bill. Bill Hatton Columns published weekly in Prince Albert Daily Herald, Prince Albert, Saskatchewan, and syndicated to other Canadian newspapers.

•Hughes, Harlan. "The Market Advisor" column. North Dakota State University Extension Service. Fargo, N.D. Columns distributed Dec. 30, 1993; Jan. 27, 1994.

•Looker, Dan. "Business." Successful Farming. Des Moines, Iowa. April, 1993.

•Magnuson, Richard. Working papers prepared for the Polish government and Volunteers in Overseas Cooperative Assistance (VOCA). Washington, D.C., and St. Paul, Minn. 1992.

•Maki, Wilbur. When is a Government Loan to a Fortune 500 Business State Economic Development? (Monograph) University of Minnesota Waite Library. St. Paul, Minn. 1993.

•Mattoon, Richard H. "Economic development policy in the 1990s – are state economic development agencies ready?" Economic Perspectives. Federal Reserve Bank of Chicago. Chicago, Ill. May/June, 1993.

•Melichar, Emanuel. Agricultural Finance: Turning the Corner on Problem Debt. (Monograph) Board of Governors, Federal Reserve System. Washington, D.C. Aug. 26, 1987.

•Miller, Cheryl. "Which Row to Hoe? An Interim Report on Alternative Directions in Agriculture." Northwest Report. Northwest Area Foundation. St. Paul, Minn. January, 1993.

•Morphew, Clark. "Church group will help Russians learn how to farm independently." St. Paul Pioneer Press. St. Paul, Minn. Jan. 15, 1993.

•Morphew. "Communal Life Rewarding for Hutterites." St. Paul Pioneer Press. Nov. 15, 1992.

•Morris, Ralph k. and Johnson, Brian D. Business Formation Options: Limited Liability Company. (Monograph) Doherty Rumble & Butler Professional Association. St. Paul, Minn. 1994.

•Morris and Johnson. Comparison of Various Forms of Doing Business in Minnesota. (Monograph) Doherty Rumble & Butler. 1994.

•The Netherlands in Brief. (Briefing papers) Ministry of Foreign Affairs. The Hague, the Netherlands. 1992.

•Pedraza, Juan Miquel. (Issues for the 1995 farm bill. Lead story.) AgWeek. Grand Forks, N.D. Jan. 3, 1994.

•Perry, Janet and Hoppe, Bob. "Off–Farm Income Plays Pivotal Role." Agricultural Outlook. Economic Research Service, U.S. Department of Agriculture. Washington, D.C. November, 1993.

•Pignatello, Nettie. "Tax reform levels playing field for schools, businesses." fedgazette. July, 1994.

•"Pig Placements and Slaughter Report(s)." Economic Research Service, U.S. Department of Agriculture. Washington, D.C. Quarterly reports. 1991, 1992, 1993, 1994.

•Pilcher, Dan. "The Third Wave of Economic Development." State Legislatures. National Council of State Legislatures. Washington, D.C. 1991.

•Pinstrup–Andersen, Per. World Food Trends and Future Food Security. (Monograph) The International Food Policy Research Institute. Washington, D.C. March, 1994.

•Public Attitudes Toward Rural America and Rural Electric Cooperatives. The Roper Organization, for the National Rural Electric Cooperative Association. Washington, D.C. June, 1992.

•Rasdal, Dave. (Tourism report.) Cedar Rapids Gazette. Cedar Rapids, Iowa. 1992.

•Redburn, Tom. "New York Pressing Bid to Lure Companies." New York Times. New York, N.Y. June 6, 1994.

•Reuwee, A. Daniel. "Family lifestyle is goal on Theis farm." Mid–Am Reporter. Mid–America Dairymen Inc. Springfield, Mo. November, 1993.

•Reuwee, A. Daniel. "Mid–Am finds opportunity in world markets." Mid–Am Reporter. November, 1993.

•Richter, Steve. "Partnerships Power Co–op Success." Cooperative Partners. Cenex / Land O'Lakes Ag Services. Inver Grove Heights. March–April, 1994.

●Rosenberg, Emily. "The New, New West: Assessing a Century of Change in the Northwest Region." Northwest Area Foundation, St. Paul, Minn. 1991.

●Salsgiver, Jackie. "Rural Development: Counties with High Percent of Ag Jobs Decline." Economic Research Service, U.S. Department of Agriculture. Washington, D.C. November, 1993.

●Saskatchewan Cooperatives: A Record of Community Development. Centre for the Study of Cooperatives. University of Saskatchewan, Saskatoon, Saskatchewan. February, 1992.

●Schrader, Lee F. and Boehlje, Michael, et. al. Summary. Indiana Agriculture 2000: A Strategic Perspective. Department of Agricultural Economics, Purdue University, and Purdue University Cooperative Extension Service. West Lafeyette, Ind. July, 1992.

●Shen, Deborah. "Food firms poised for battle." The Free China Journal. July 1, 1994. Taipei, Taiwan.

●Stern, William M. "Land O' Low Returns." Forbes Magazine. Aug. 15, 1994.

●Testa, William. "Trends and prospects for rural manufacturing." Economic Perspectives. Federal Reserve Bank of Chicago. Chicago, Ill. March/April, 1993.

●Thompson, David. "Rochdale Revisited." Cooperative Business Journal. Washington, D.C. October, 1993.

●"2020 Vision Initiative Launched." News & Views newsletter. International Food Policy Research Institute. Washington, D.C. June, 1994.

●Third World: Customers or Competitors? E.A. Jaenke & Associates. (Monograph.) Washington, D.C. August, 1987.

●Veblen, Tom C. The Politics of Trade in Foodstuffs. (Monograph) Enterprise Consulting Inc. Washington, D.C. Oct. 18, 1993.

●von Witzke, Harald, Chichon, Janusz; and Hausner, Ulrich. Economic Reforms in Poland: Implications for Agriculture. (Monograph) Center for International Food and Agriculture Policy, University of Minnesota. Minneapolis, Minn. 1993.

●Watts, Thomas G. and Lee, Steven H. "Reviving the Great Plains // Pioneer Spirit Lives as People Fight to Save Communities." Dallas Morning News. Dallas, Tex. June 2, 1991.

●Webb, Tom. (Wireservice dispatch.) Knight–Ridder Newspapers. Washington, D.C. Feb. 17, 1994.

●Willkommen. Amana Convention and Visitors Bureau. Amana, Iowa. Summer edition, 1993.

●Worthington, Rogers. "Sociologists say some small towns ought to give up the ghost." Chicago Tribune. Chicago, Ill. Dec. 26, 1989.

INDEX

A

A & W Brands Inc. • 53

a piece of the rock, We Indians got... • 95

A Thousand Acres • 24

A.C. Toepfer International • 50, 133

Aberdeen American–News • 72, 244, 260

Abourezk, James, Sen. • 76, 97, 258

Adams, Paul • 259

added value: via NFO marketing services • 231

Addis Ababa • 70

Advance Seed • 52

Aeilts, David • 235

Afghanistan • 73, 74

Africa • 37, 104, 117, 203

African, Somali and East African immigrants • 194

Agassiz Pork Cooperative • 247

Ag–Chem Equipment • 194

Agnew, Spiro, Vice President • 153, 154

Ag–Nomics Inc. consulting service in New Brighton, Minnesota • 188

agrarian myth (myths) • 24, 25, 27, 32, 33, 41, 123

Agri Grain Marketing(AGM) • 237

Agri Industries, cooperative, West Des Moines, IA: partnership with Cargill • 237

Agri News • 259

AgriBank • 160, 174; St. Paul, MN cooperative • 223

AgriBank FCS • 174

Agricultural Cooperative Development International(ACDI) • 168, 171

agriculture: divine right of • 62

agriculture, animal, structure changing • 233

AgriVisions magazine • 4, 245

Agway • 128, 130, 134, 172, 235

AgWeek magazine • 32, 45, 259

Alabama • 190, 205, 207

Alabama, Vance • 204

Alaska • 75

Albania • 167, 168

Albert Lea, Minnesota • 79

Alliance for Progress program • 71

all–weather political speech • 34

Almen, Jan • 260

alternative crops • 125

Alternative Energy Resources Organization, Helena, Montana • 125

Amana: Colonies • 175, 179, 180, 182, 187, 190; Colonies, the Great Change • 180; Furniture Shop • 181; Holiday Inn • 181; Meat Markets • 181; Refrigeration Co. • 181; Society • 180, 181; Society, 700 shareholders • 180; villages, yearly tourist take $30 million • 179; Woolen Mill • 181

Amarillo Globe News • 260

Amato, Joseph A., Dr. • 218, 258

Amato–Meyer book • 219, 225

Amber Milling • 133

Amdahl, Burgee • 160

America's raw material industries the shock absorber • 75

American Agriculture Movement (AAM) • 28, 29, 31, 81

American Airlines • 206

American Crystal Sugar • 135, 147, 228, 258; secondary stock offering via expanding acerage • 149

American Farm Bureau Federation • 31, 168

American Farmland Trust • 113

American Indian Chamber of Commerce • 188

American Prairie Foods • 246

American soybean trade embargoes: Japanese reaction to • 26

Amish, Old Order Amish • 178, 182, 184, 185, 186, 187, 190

Amstutz, Daniel, Ambassador • 212

273

horizontal farm expansion (acreage growth): need for reduced by co-ops • 223

Hormel Foods company • 78, 194

hormones, growth: cows not treated with • 220

Houck, James • 171, 259

Howe, Ray • 259

Hubert H. Humphrey Institute of Public Affairs • 17, 44, 76, 111, 240

Hughes, Harlan • 232

Humboldt University(Germany) • 258

Humphrey, Hubert H. • 156, 158, 170

Hungary • 156, 167

Hutterite: community near Brookings, SD • 184

Hutterites • 183, 184; building industrial communities for their followers • 184; own Community Playtings brand name • 185; start industrial community, Dexter, Minnesota • 184

I

IBM Corp. • 205

IBP Co. • 194, 211

Idaho • 34, 238; Council of Cooperatives • 191

Illinois • 62, 114, 207; Quad Cities • 211

Imperial Holly sugar company • 148

Independent–Republicans(IR) • 34

India • 158; New Delhi • 173

Indian Farmers Fertiliser Cooperative Ltd. • 173

Indiana • 207; Economic Development Council • 207

Indianapolis Star • 260

induced technological change • 69

inflation, jawboning • 201

Ingersoll, Bruce • 260

Institute for Food Studies & Agroindustrial Development, Copenhagen • 16

interdenominational assistance program • 163

interest rate buydown programs • 88

International Cooperative Development Association • 168

International Cooperative Petroleum Association • 173

International Food Policy Research Institute • 197

International Herald Tribune • 260

International Monetary Fund • 28

International Multifoods Co. • 131, 150

International Wheat Organization(IWO) • 212; model for OPEC • 200

Inter–Religious Task Force • 182

Iowa • 61, 109, 175, 180, 184, 209; Amana Colonies • 179; Center for Agricultural and Rural Development, Iowa State University • 161; Ebenezer community begins migrating to, 1855 • See Amana; explores ways to combine manufacturing base with ag strengths • 238; Iowa Pork Producers Association • 138; Iowa State University • 57, 163, 223, 226, 237, 259; new rural America rising in ruins of • 217; northwestern • 185; precinct caucuses • 27; Quad Cities • 211

Iowa Farmer Today • 260

Iowa River Valley • 180

isolationism • 156, 157, 158, 170

Israel • 116

Issues Strategies Group company • 97

Italy • 52, 85, 209

J

J.P. Morgan company • 206

Jackson, Andrew, President • 27

Jackson, Henry 'Scoop', Sen. • 158

Jackson, Jesse, the Rev. • 23

Jamaica • 155, 156

James Horvath, President, Golden Growers • 14

James River Farmers Elevator • 4, 235

James, Carol • 260

North American Bison Cooperative • 247

North American Free Trade Agreement • 137

North Carolina • 40, 205, 206, 211; lost biggest smokestacks it ever chased • 204; smokestack chasing give-a-ways no match for Alabama's • 207; swine multipliers • 237

North Central Cattle Feeder Cooperative • 247

North Central Distributing cooperative • 135

North Central Sugar Marketing • 135

North Dakota • 13, 14, 35, 40, 61, 75, 157, 158; A Bicentennial History • 45; Agricultural Products Utilization Commission, special state agency • 226; Carrington • 225; Coordinating Council on Cooperatives • 225, 245; new rural America rising in ruins of • 217; State University • 149

North Dakota Open • 95

North Dakota State University • 232

North Pacific Grain Growers cooperative • 133

Northern Corn Processors Inc. • 16, 148

Northern Europe • 14, 34, 37, 39, 52, 229, 236

Northern Integrated Pork Cooperative • 247

Northern Plains Organic Cooperative • 225, 247

Northern Plains politics... in context of wheat economy • 157

Northern Plains Sustainable Agriculture Society, Wales, North Dakota • 125

Northwest Airlines • 116, 154

Northwest Area Foundation • 213

Norway • 86

Nourse I cooperatives • 146

Nourse II cooperatives • 146

Nyquist, Robert • 148

O

O'Day, John, vice president of AgriBank • 257

O'Grady, Don • 260

O'Reilly, Dani • 259

Oakwood Bruderhof • 184

Ocean Spray cooperative • 130; specialty, in Wisconsin, Massachusetts • 229

Ochs, John • 26

Oglala Sioux • 94, 189

Ohio • 54, 207, 210; Janesville • 120

Oklahoma • 40, 160

Olive Garden restaurants • 131·

Olivia, MN • 221

Olson, Allen I., former North Dakota Gov., head of Independent Community Bankers of Minnesota • 35, 96, 224

Olson, Floyd B., former Minnesota governor (1932) • 82

Olson, Van • 96

Olsztyn University of Agriculture and Technology (Poland) • 166

Omaha World Herald • 260

Ontario • 55, 98

Oregon • 36, 39, 210, 238

Organic Growers & Buyers Association, New Brighton, Minnesota • 125

Organization for Economic Cooperation and Development(OECD) • 45

Organization of Petroleum Exporting Countries (OPEC) • 77, 78, 79, 199, 200, 212, 213

Orr, Richard • 259

Orval Kent Foods • 144

Osborne, David • 197

Oukrop, Al • 226

Oxford University (U.K.) • 156

P

Paarlberg, Robert • 171

Pacific Star de Occidente(Mexico) • 53

S

292

Y

Z

A field trip.
Author Lee Egerstrom with
daughters Kari, Kirsten and Marisa.

The Computer Infobase®™ Edition of "Make No Small Plans" is available directly from:

Lone Oak Press, Ltd.
304 11th Avenue SE
Rochester, MN 55904

Send $16.95 + $2.00 shipping. The Infobase Edition is issued on 3.5" High Density (1.44mb) disks, is indexed for text search (Boolean and complex multiple phrase) and retrieval, allows annotation, custom notation and the full capabilities of its Folio Views®™Infobase software engine. The Infobase Edition contains, in addition to the full test of "Make No Small Plans", hundreds of pages of supporting documentation including theses, white papers, research studies and planning reports.